ALL WE WANT IS THE EARTH

Land, Labour and Movements
Beyond Environmentalism

Patrick Bresnihan and Naomi Millner

BRISTOL
UNIVERSITY
PRESS

First published in Great Britain in 2023 by

Bristol University Press
University of Bristol
1–9 Old Park Hill
Bristol
BS2 8BB
UK
t: +44 (0)117 374 6645
e: bup-info@bristol.ac.uk

Details of international sales and distribution partners are available at bristoluniversitypress.co.uk

© Bristol University Press 2023

British Library Cataloguing in Publication Data
A catalogue record for this book is available from the British Library

ISBN 978-1-5292-1832-9 hardcover
ISBN 978-1-5292-1833-6 paperback
ISBN 978-1-5292-1834-3 ePub
ISBN 978-1-5292-1835-0 ePdf

The right of Patrick Bresnihan and Naomi Millner to be identified as authors of this work has been asserted by them in accordance with the Copyright, Designs and Patents Act 1988.

All rights reserved: no part of this publication may be reproduced, stored in a retrieval system, or transmitted in any form or by any means, electronic, mechanical, photocopying, recording, or otherwise without the prior permission of Bristol University Press.

Every reasonable effort has been made to obtain permission to reproduce copyrighted material. If, however, anyone knows of an oversight, please contact the publisher.

The statements and opinions contained within this publication are solely those of the authors and not of the University of Bristol or Bristol University Press. The University of Bristol and Bristol University Press disclaim responsibility for any injury to persons or property resulting from any material published in this publication.

Bristol University Press works to counter discrimination on grounds of gender, race, disability, age and sexuality.

Cover design: Andrew Corbett
Front cover image: V'cenza Cirefice
Bristol University Press uses environmentally responsible print partners.
Printed and bound in Great Britain by CPI Group (UK) Ltd, Croydon, CR0 4YY

For Ted

The cover illustration, by V'cenza Cirefice, is based on Rosebay willowherb. Used by people across the world, for medicine, food, weaving, and making tea from the leaves. A meadow plant that likes to grow on disturbed ground, on 'waste-lands', the edges of construction sites, roads, mines. Known as fireweed in North America because it is the first to grow after a wildfire, the seeds stay in the soil ready to populate freshly disturbed land. You also find it on reclaimed bogs. It cares for and repairs the soils before dying back and allowing other species to flourish. Each plant produces up to 80,000 seeds with fluffy sails that allow them to travel huge distances and to grow very quickly. Like the afterlives of the movements in this book, this plant nurtures the others that come after it in all their different forms.

Contents

List of Figures and Boxes		vi
About the Authors		viii
Acknowledgements		ix
1	Introduction: Beyond Modern Environmentalism	1
2	Suburb, Field, Laboratory: Recomposing Geographies of Early Environmentalism	25
	First Interlude: Green and White Dreams	44
3	Revolt Against One-Worldism: Radical Claims on Land and Work Post-1968	54
	Second Interlude: Planetary Icons	75
4	The Right to Subsist: Transnational Commons Against the Enclosure of Environments and Environmentalism	81
	Third Interlude: Witnessing in the Global Resonance Machine	102
5	Earth Politics: Disagreement and Emergent Indigeneity in the So-Called Anthropocene	105
	Fourth Interlude: Making Things Resonate	124
6	Conclusion: Resonance Beyond Environmentalism	128
Coda: Afterlives		141
Notes		145
References		157
Index		176

List of Figures and Boxes

Figures

1	Du Pont Cellophane advert, 1954	45
2	34th anniversary of the Bhopal tragedy, 3 December 2018	46
3	Advertisement for chemicals produced by Union Carbide, c. 1954	47
4	'DDT is good for me-e-e!', *Time Magazine*, 30 July 1947	48
5	Flit advertisement, 1930–1940, Dr. Seuss Advertising Artwork	49
6	Protests against crop dusting, date unknown	50
7	British advertisement for Shell petrol, featuring wildlife in Wales painted by Maurice Wilson, 1955	51
8	Greenpeace activists protest at a Shell gasoline station in San Francisco, California, 1995	52
9	Ogoni Solidarity Ireland, County Mayo, 1995	53
10	'The Psalter Map', c. 13th–15th century	76
11	Persian cosmography map, '*Kharidat al-'Aja'ib wa Faridat al-Ghara'ib*' [The Pearl of Wonders and the Uniqueness of Things Strange], c. 1525	77
12	'Earthrise', 1968	78
13	'Blue Marble', 1972	78
14	Juan de Santa Cruz Pachacuti Yamqui Salcamaygua: Cosmology of the Inca, 1613	79
15	5th Summit of the Non-Aligned Movement Logo, 1976	80
16	Poem written by Ken Saro-Wiwa, 1995	93

Boxes

1	Denaturalising environmentalism	7
2	Provincialising environmentalism	15
3	Pesticides in war and nature	30

4	Environmental racism and Black ecology	36
5	Zerowork and the trans-Atlantic connection	64
6	Witnessing practices	88
7	*Wild Relatives* (2017) by Jumana Manna: worlding seed conservation from below	111

About the Authors

Patrick Bresnihan is Lecturer in the Department of Geography at Maynooth University. He works across the interdisciplinary fields of political ecology, science and technology studies, and environmental humanities. His research and activism focuses on related issues of land, water and energy in Ireland. He is the author of *Transforming the fisheries: Neoliberalism, nature and the commons* (University of Nebraska Press, 2016).

Naomi Millner is Senior Lecturer in Human Geography, Geographical Sciences, at the University of Bristol. She is a political geographer and political ecologist who explores the knowledge politics surrounding the making and management of global 'environments' in the context of changing global agendas for sustainability and changing terrains of conflict. Alongside her academic work Naomi is active in the Bristol Hospitality Network, a solidarity housing network for asylum-seekers, and is a keen community gardener.

Acknowledgements

We didn't know it at the time, but this book began when we first met at a conference in Nottingham in April 2011. Over the last ten years or so, our intellectual collaboration has been inseparable from our friendship. The people we are most indebted to are also people we consider friends and allies, as well as intellectual inspirations and collaborators. Some of these people are closer than others, but all have influenced us through the generosity of their thinking, willingness to support us, and commitment to the cause of social and environmental justice.

We are inspired by the work of many scholars and writers. This goes beyond the referencing of individual texts. Peter Linebaugh and John Holloway are passionate writers who have always reminded us of the power of words. Wendy Larner and Eamon Slater have been important mentors. Christian Parenti, Sian Sullivan, Jason W. Moore, Luigi Pellizzoni, Katharine Gibson, Rosaleen Duffy, Stefania Barca and Jennifer Gabrys are some of the most important thinkers on environmental politics today and have always had the time to talk and be supportive. A special acknowledgement is reserved for Dimitris Papadopoulos and Maria Puig de la Bellacasa who are bedrocks and always inspire us with their work as well as their convivial approach to scholarship.

Naomi is indebted to networks of friends and colleagues in Colombia, Guatemala, Peru and Mexico who have been coming together to explore political ecologies of nature-technologies and the making of worlds; also to those collaborating on interdisciplinary projects in Colombia and Guatemala, including academics Dunia Urrego and Toby Pennington, and the BioResilience team; the political ecology reading group involving wonderful master's and PhD students, and community researchers; the activists and communities of Serranía de las Quinchas, Monquentiva and Flores, and the interdisciplinary drone ecologies team, including Libby Lunstrum, Chris Sandbrook, Jaime Paneque-Galvez, Nico Vargas Ramírez, Anna Jackman, Andy Cunliffe, Mara Mulero, Serge Wich, Charis Enns, Brock Bersaglio, Ben Newport and Trishant Simlai. Also all of those involved in Haciendo Mundos in Bogota in 2021 and the associated retreat, especially Astrid Ulloa, Diana Ojeda and Arturo Escobar.

We are grateful for conversations and inspiration from Roz Corbett, Mama D Ujuaje, Sarah-Poppy Jackson, Ian Fitzpatrick, Ben Moss, Marika Rose,

Nate Millington and others active in the UK on questions of ecological reparations, food sovereignty, land justice, anti-racism and anti-colonialism. In Ireland, Patrick Brodie, Sive Bresnihan, Sinead Mercier and V'Cenza Cirefice, whose critical work on environmental politics, extractivism and decolonisation is an important reference point for this book. We would like to acknowledge the many important exchanges we've had over the years through ENTITLE, POLLEN and Degrowth political ecology networks, particularly with Gustavo Garcia, Emanuele Leonardi, Panagiota Kotsila and Melissa Garcia-Lamarca. Also, Manuela Zechner and Bue Rübner Hansen whose work with migrant, ecofeminist and ecological social movements has helped our own thinking for this book.

Thanks to our colleagues in the Geography departments in Trinity College Dublin, Maynooth University and Bristol University, especially Cian O'Callaghan, Phil Lawton, James Merricks-White, Arielle Hesse, Rory Rowan, Oona Morrow, Kathleen Stokes, Karen Till, Gerry Kearns, Mary Gilmartin, Louise Fitzgerald, Negar Behzadi, Franklin Ginn, Mark Jackson, Merle Patchett, Joe Gerlach and Maria Fannin, with whom important conversations were had during the shaping of this book. PhD and postdoctoral researchers we have both worked with and supervised, who also at important moments, are also our collaborators and motivate us continually.

Our friendship and intellectual collaboration would not have been possible without our involvement in the Authority Research Network (ARN). Formed out of a desire to combine critical and collective thinking, reading, writing, eating and drinking together, we have spent some of our most productive and pleasurable times together with the ARN – in Ireland, England, the US and Brazil. Without the ten years of camaraderie, humour and conversations with Tehs Noorani, Leila Dawney, Claire Blencowe, Jules Brigstocke, Aecio Amaral and Sam Kirwan, this book wouldn't have happened.

As a graduate student, Paddy was lucky to make friends with Alex Dubilet, Tom Stammers, Ross Perlin, Mick Byrne and Alessandro Zagato, all of whom changed how he saw the world. The same can be said for Marcela Olivera and the formative time he spent in Cochabamba, Bolivia. Naomi would like to acknowledge Mónica Amador in particular – a close collaborator who began in the BioResilience project as a post-doc, but has gone on to become a co-author, co-conspirator and wonderful friend.

We've both experienced challenges during the writing of this book reminding us of the value of brothers, sisters, mums, dads, cousins, nieces, nephews and the extended network of family and friends. Rachel O'Dwyer is the best friend, partner and mother a person could wish for. Ted, you are a superhero.

1

Introduction: Beyond Modern Environmentalism

Between 31 October and 13 November 2021, the 26th United Nations (UN) Climate Change Conference, known as COP (Conference of the Parties)-26, was held in Glasgow, Scotland. Despite pressure to find ways forward through the climate crisis and reduce emissions, the Conference was dismissed by many as a corporatised, greenwashed 'spectacle': an event that reproduced colonial inequalities by shutting out Indigenous leaders while focusing on solutions that would enable business as usual in the richer countries of the Global North. While world leaders gathered inside the walls of the SEC Centre, Indigenous leaders denied entrance gathered across the river Clyde to commemorate activists killed for trying to protect the planet from corporate greed and government inaction (Lakhani 2021). The main summit focused on achieving 'net-zero targets', which Indigenous movements, peasant organisations and land coalitions call out for being false climate solutions promoting further land grabs and bio-cultural destruction. Net-zero solutions were accused, here, of reproducing the dynamics of colonialism and extractivism that produced the environmental crisis in the first place, incentivising carbon capture markets through mass reforestation, biofuels and new technologies.

If you have chosen to read this book, you are probably familiar with critiques of COP and other global environmental summits. This is not a new story, and we could have opened with a similar account of previous COP meetings. As the popularity of the recent satirical film *Don't Look Up* (2021) made clear, many people know that sclerotic political institutions and powerful global corporations are intensifying, rather than halting, planetary ecological breakdown. Yet the problem is more deeply rooted and relates to the longer histories of modern environmentalism. The premise of this book is that digging into these histories will help us understand how to move beyond the blockages and blindspots of modern environmentalism to make more globally just ecological futures possible.

This look into the past also enables us to appreciate that the consolidation of global environmental management since the 1990s is not itself a new phenomenon. Rather it marks a continuation of inherited assumptions and tendencies throughout environmentalism's history that reproduce what and who is counted as worthy of protection, and what and who isn't. Examining how this situation has come about can help us understand – and perhaps avoid – making social and political dispossession part of the nature of environmental protection.

Over the past 60 years, the term 'environmentalism' has come to refer to the coordination of actions and activities for the protection of 'the environment' – referring to ecological and nonhuman systems, landscapes and the diverse nonhuman species that compose them, understood to be threatened by the scope and pace of human industrialisation. It has taken myriad forms of labour – scientific, activist campaigning, storytelling, direct action – to make these concerns matter to wider audiences and enact changes at the level of law, policy and public awareness. This book is about what has been produced as a result: modern environmentalism. For us, modern environmentalism refers to a way of seeing and thinking about the environment that extends to a broad field of actors and actions – including environmental activists, environmental non-governmental organisations (NGOs), journalists and academics like us working on 'environmental' topics. By unravelling the distinct ways of seeing, knowing and naming that comprise modern environmentalism, we will enact disagreement with its basic terms, illustrating how 'environmental' concerns can participate in reproducing inequalities and perpetuating socio-ecological harms.

Our intention in writing this book is not to undermine the individuals or the attitudes of care and commitment embodied in and through practices of environmentalism. Rather, we seek to revisit what is at stake in the ways environments are named, framed and fought for. We can see the implications through the way *Don't Look Up* tells its story, portraying the familiar trope of the hero scientist coming up against corrupt politicians and vested interests. More than anything, the film represents a mood of general cynicism about the possibility of actual change, even as it marks a powerful (and widely shared) critique of the paradoxes of global environmental governance. We encounter this mood in our teaching, as many of our students are sceptical about the claims and promises of policy makers and companies, but feel disempowered when thinking how this might be remedied. There is a quiet, often individualised, discontent at large that has few political avenues for articulation. But we might ask, where are the social movements in the film? Where are the alternative ways of telling the story, building on long-standing forms of anti-racist, anti-colonial, feminist critique? Revisiting the ways that narratives are told in environmentalist thought, we suggest that all is not lost. At the edges of things, there are different ways of seeing, organising and thinking environments. We turn our attention to these edges.

This brings us to the core motivation for this book. While the big picture does not look hopeful, there are always other places to look – events, collective energies, objections that gather around concerns that don't always come under the name 'environmentalism', and yet are crucially connected to the struggle for more just ecological futures. By looking at social movements, thinkers and claims beyond the mainstream, from the 1960s until recent times, we want to explore fresh ways of linking ecological, justice and equity issues. These movements and thinkers did not always consider themselves 'environmental' and are often counted outside the history of environmentalism retrospectively. However, our premise is that these wider spheres of objection, disagreement and resonance have much to offer in addressing problems of ecological damage and questions of reparative futures – and in helping us call out false solutions. With many others, we want to consider whether attention to movements happening adjacent to, but beyond modern environmentalism, can help us compose the collective power we need to confront climate and ecological breakdown.

In this book we make three key interventions. First, *we unpick and challenge key assumptions about the nature of environmental protection consolidated within national and global policies, institutions and public discourse over the past 60 years.* We do this by following the making of modern environmentalism in key historical periods. Leaning on older aesthetic traditions and shaped by changing political-economic pressures, environments have been imagined and regulated as places without humans, diverse ways of living and political histories. Environments come to be understood as the *backdrop* to life: passive matter that needs to be protected, at all costs, by powerful and charismatic voices. This pushes struggles relating to human labour and human–environmental relations and access to land out of the picture, as if these were separate from environmental concerns. The role of political and economic systems wired for profit maximisation in shaping environments are rendered invisible, as the emphasis falls on 'saving' nonhuman species. More recently, such forms of protection have resulted in highly technical and bureaucratic assemblages of computerised models and market mechanisms. Here environmentalism lends itself as a convenient banner to other economic and political agendas. If we act in the name of the environment, who can disagree?

These tendencies, political and aesthetic, explain why social movements have frequently organised for ecological concerns *without* using the term environment. This is our second intervention: *many social movements have surfaced over the past 60 years that have not acted in the name of the environment, or described themselves as environmental, and yet are invested in campaigns and projects that are about care for and protection of the environment.* The core threads of care for ecologies, defence of sustainable livelihoods, and cultivation of ways of life where humans and nonhuman thrive together are much older and wider than what we have come to know as environmentalism. Indeed, these threads have featured centrally in movements that centred on decolonisation, land, environmental health and resistance to large-scale development projects. These movements have given rise to internationalist networks organised under other names – agroecology, food sovereignty, territorial rights, Indigenous sovereignty, earth politics – connecting concerns of ecological damage/repair with political and social demands side-lined in many environmental projects. This indicates that there is a fundamental disagreement at the heart of environmental politics, centring on what counts as 'environmental' and who gets to determine that.

By linking these struggles *beyond* environmentalism with environmentalism's histories and dilemmas, we illuminate chinks of hopefulness within a landscape marked by ecological devastation and failures to halt the damage. Our third intervention then is directed at those who are deeply concerned about the undeniable trajectories of climate and biodiversity breakdown but are unclear where to invest that concern. *Building a viable planetary future will involve becoming part of struggles that may at first seem unrelated to environmentalism: struggles over land, labour and decolonisation.* The histories and figures recounted in this book are not presented as a dead past but as unfinished projects to be reclaimed as part of our present struggle against capitalist and colonial systems of exploitation and ecological harm. We argue that it is not possible to repair the (environmental) damage, without confronting these systems.

In the remainder of this introductory chapter, we expand and unpack some of the major arguments in the book, as well as the histories and literatures we draw on to make the three outlined interventions. There is not enough space to talk about all the literatures we are influenced by, however, we use endnotes extensively to acknowledge our sources, provide further detail and suggest additional readings. In the next section we revisit the history and formation of modern environmentalism, exploring some of its key assumptions. In the third section we emphasise three important blindspots that have been configured by this specific history, proposing these as productive entry points for a counter-history of modern environmentalism. Then, in the fourth section we use the title of the book to clarify what this project will, and will not, set out to do, as well as situating our own research and commitments as authors. Finally we explain the structure of the book, and how we hope it will be read.

A brief history of modern environmentalism

The birth of a movement: environmentalism since the 1960s

Modern environmentalism emerged in the 1960s in response to industrial technologies and forms of production that were on a trajectory to make the earth unliveable for humans (Buell 2004). Both the problem identified here, and the prompt for reparative action, rest on the powerful idea that humans can transform their own milieu, for better and for worse; modern environmentalism thus hinges on conceptualising the place and power of human agency.

Of course, this idea was not totally new, but conceptions of the (planetary) scale of the damage, and the development of new fields of environmental science, began developing rapidly from the mid-20th century. The birth of the environment, and thus environmentalism, is inseparable from these emerging ways of knowing and representing a world changed by human action (Warde et al 2018).[1] The fields of knowledge and intervention that resulted are not limited to science, but extend to art and culture – a wider aesthetic regime[2] that seeks to make sense of, and regulate, human–nonhuman interactions. This book traces some of the evolution of these ways of knowing and representing environments, drawing on the more detailed work of historians who have made such shifts central to their work (see Box 1).

The major environmental concern in the 1950s and 1960s was the existential threat posed by nuclear war. The possibility of planetary-scale devastation was channelled through popular culture and reinforced through advances in scientific research often afforded by the very technologies that threatened planetary survival (see Masco 2010). Space travel made possible the 'earthrise' pictures taken by Bill Anders in 1968 (see Second Interlude) and the vision of a fragile planet suspended in space (Jasanoff 2004). Space travel also triggered interest in ecological and biospheric design and extra-planetary exploration, directly informing the interest in manipulating earth systems (Anker 2005).

The threats posed by nuclear testing and the toxic risks associated with chemical pesticides also spawned citizen science movements and public interest in environmental matters across this period. As we explore in Chapter 2, Rachel Carson's stirring and poetic book on the risks of chemical pesticides is still considered by many to be the work that launched the modern environmental movement (Lear 1998). *Silent Spring* (Carson 2002 [1962]) articulates a central contradiction of this period: the promise of the post-war boom was based on untested technologies that posed unprecedented risks to present and future generations. Carson's book captures the traffic between new scientific thinking (Carson was a trained marine biologist) with a popular culture of concern and fear around the uncertain impacts of new technologies on vulnerable and complex environments. Such mediating figures played a key role in galvanising civil society for change. However, as we will show, other collectives working to transform public perceptions of pesticides are less well-remembered. Environmentalism has developed a selective memory that highlights figures, events and forms of objection, while downplaying others.

One of the central tenets of modern environmentalism as it has developed since the 1960s is faith in the ability of science and technology to manage environmental problems (Boudia and Jas 2014). This was evident at a global, as well as a national, level. The 1972 *Limits to growth* report applied computer modelling at a global scale for the first time. Its findings provided convincing evidence of the unsustainability of current trajectories of development, based on the projection and correlation of major variables such as population, agricultural production and pollution. Despite criticism of its crude methodology, this modelling experiment set in motion trends for informing governance at the planetary scale through the elaboration of large datasets, something that takes on clearer institutional form with the establishment of the Intergovernmental Panel on Climate Change (IPCC) in 1988.

As ecological problems became framed as complex, dynamic and global, we also see a convergence between environmental management, networked technologies and markets (Walker and Cooper 2011; Leonardi 2012).[3] Rapidly evolving information and communication technologies, particularly the connectivity and networking possibilities of the internet, has facilitated emerging ways of seeing environments and their structuring outcomes (Gabrys 2016). Modern environmentalism consolidated during the 1980s and 1990s as a globalised regime to measure, monetise and manage distinct pollutants, flows of energy and ecosystems, steadily pushing from view other framings of environmental problems and their causes. Thus we see the emergence of environmental organisations from the 1990s onwards that also tackle issues of biodiversity loss and deforestation through the application of models without necessarily consulting local people and organisations. Meanwhile, the continued innovation of eco-markets permits rapid financialised transactions between buyers and sellers, while shaping the sorts of environmental 'products' that can be marketed and traded via electronic exchanges. Such abstractions are problematic because places – and people's relationship to them – are highly particular, and are often informed by longer histories.

In this book we highlight how 'the environment' became established as a kind of hidden universal through such steps – a concept that seems to be about protecting what matters for everyone, and yet is often about protecting a particular version of human–nature relations. Crucially, establishing global and managerial solutions to environmental problems has been achieved at the cost of heeding more radical critiques of industrial technology and capitalist political economy. Nevertheless, attending to the making of environmentalism allows us to see that at each of these steps, fresh objections and efforts to do environmental politics otherwise have emerged. It is important to emphasise the hope that can be found by looking into the shadows of environmentalism: by connecting with the energies of social movements we can expect to find forms of collective action and analysis that do not reproduce the same blindspots.

Box 1: Denaturalising environmentalism

What is now popularly called the environmental humanities originates in part from the new subdiscipline of environmental history, associated most closely with the work of US-based scholars Donald Worster, William Cronon, Richard White, Denis Cosgrove and, less well-known, Frieda Knobloch. Largely focusing on the US, this work demonstrates the effects of settler-colonialism, industrialisation and urbanisation since the 19th century not only on landscapes, ecologies and society, but on an emerging ideology of wilderness and pristine nature. The extent to which these romantic and highly problematic ideas of nature continue to inform modern environmentalism is a key theme in the influential edited collection *Uncommon ground*, the product of a semester-long seminar at the University of California, Irvine held in 1994 involving, among others, Donna Haraway and Candace Slater. As the organiser of the seminar, historian William Cronon, wrote in the introduction to the book: '[I]f we hope for an environmentalism capable of explaining why people use and abuse the earth as they do, then the nature we study must become less natural and more cultural' (Cronon 1996b: 36).

One of the participants in the University of California, Irvine seminar in 1994 was Donna Haraway. A feminist philosopher of science, Haraway's work seeks to unsettle the inherited patriarchal, racist and capitalist norms imbued into dominant (or 'universal') science. Haraway explicitly counters accusations that this critical labour undermines possibilities for truth or objectivity – a criticism launched at Cronon and contributors to *Uncommon ground* at a time when environmental organisations were vulnerable to funding cuts and attacks from right-wing, conservative elements (which we see returning with a vengeance in the form of 'post-truth' discourses). With other feminist scholars like Sandra Harding (1992), Haraway argued for situated knowledge which aims for objectivity that is unavoidably partial and always open towards building better relationships with the world. Rather than getting lost in debates over universalism versus relativism, the challenge for Haraway (and many critical scholars and activists who affirm alternative systems of knowledge and practice) was to understand how authority and expertise might be reconstituted in more plural, egalitarian and disruptive ways (Brigstocke et al 2021).

It is not surprising that important scholarship interrogating the cultural assumptions of modern environmentalism emerged from the US, which has been at the forefront of environmental transformations driven by the globalisation of new chemical and biotechnologies since the Second World War. At the same time, the US is also commonly seen as the birthplace of modern environmentalism. As we argue in this book, the coincidence of these developments is not accidental. Environmentalism was very much a product of US-based progress thinking, as well as US-based wilderness imaginaries. The early work of environmental humanities scholars sought to connect these, and longer, histories, showing how core assumptions of the mainstream environmental movement were ideologically aligned with the very systems they sought to challenge (Braun 2002).

More recent work in environmental history focuses on the complex relationships between environmental expertise, social movements, ecologies, chemicals, infrastructures and capitalist political economy – what Michelle Murphy (2008) has called 'chemical regimes of living'. Murphy's book on the politics of 'sick-building syndrome' connects histories of office architecture, industrial chemistry, feminist activism, public health expertise and environmental science, illuminating just how enmeshed different spheres of social, technological and ecological activity have become, and, as a result, the kinds of open methods we need to examine these. Linda Nash's wonderful book *Inescapable ecologies* (2007) is another example, narrating a non-linear, critical history of Southern California that combines histories of capital and colonisation through the reworking of relations between body and environment, health and place.

Other key scholars in the field of environmental history/science and technology studies are Christophe Bonneuil and Jean-Baptiste Fressoz, Paul Edwards, Gabrielle Hecht, Jennifer Gabrys, Max Liboiron, Gregg Mitman, Soraya Boudia and Nathalie Jas. Through their careful and rigorous work, these scholars show (rather than just state) that science and technology are value-laden and mediate social practices in complex ways. Science and technology are part of social systems and ecologies, and are also contested, appropriated and recuperated in ways that challenge and remake worlds.

Such insights point to the importance of *aesthetics* in the making of modern environments and environmentalism. By aesthetics, we mean selective ways of seeing and sensing the world that illuminate some relations while excluding others. In this book, we connect such tensions and exclusions with the cultural production of advertising, literature, visual art and media representation, which have systematically reinforced environments as particular 'scenes'. Consolidating such scenes in historical moments foregrounds some actors, human and nonhuman, while backgrounding others. In the following chapters, we focus on the second half of the 20th century to explore how and why this aesthetic production took place, and what the consequences were. First, however, we summarise three important blindspots (([anti-]colonial geographies; labour/social reproduction; lived, embodied realities) in modern environmentalism as it consolidated from the 1960s until today, which form crucial themes throughout the book.

Blindspots in modern environmentalism

(Anti-)colonial geographies

If modern environmentalism is bound up with a belief in (some) people's ability to manage the environment, it is also wedded to ideas of protecting nature from (other) people. This paradox – which sets the need for human intervention against the need to stop human intervention – hinges on assumptions about the separation of Nature and Society which has a well-documented history within Western thought.[4] While environmentalism emerged in the 1960s, it inherits cultural assumptions from longer histories and aesthetic traditions.

One of these aesthetic traditions, the 'wilderness' imagination, was central in movements to establish the early National Parks. Wildernesses are physical environments that humans did not create, and yet, the idealisation of wilderness as a spiritual sanctuary apart from humans is 'quite profoundly a human creation' (Cronon 1996a: 69). In the context of the North American frontiers in the late 19th century, new National Park sites demarcated pilgrimage destinations for white settlers from cities like San Francisco and Sacramento, not for the First Nations peoples who inhabited them (Spence 1999). This pull 'out into nature' was nourished by the promise of spiritual transformation in Romanticist art and painting, which portrayed humans as being dwarfed against backdrops of luminous and magnificent Nature. Such spatial imaginations allowed visitors to appreciate the wonders of nature without needing to confront the forcible removal of people – the original inhabitants of National Parks, for example, such as the Blackfeet people in Glacier National Park, often had rights to hunt and fish erased, without compensation, or were forcibly removed. Wilderness concepts conveniently obscure how First Nations people were forcibly removed from those spaces. They also obscured the ideas of 'creating a master race' embedded in conservation aesthetics – early conservationists, such as Madison Grant, were key players in major eugenics conferences (Spiro 2009).[5] Through such blurring, wild spaces become inseparable from quintessential American-ness (or Britishness), masculinity (being rugged, a real man) and elitism, for example, for big game hunting in the Rockies and in the Scottish Highlands. Political ecologists, along with literary scholars, cultural historians and critical theorists, revisit this wilderness thinking to show us that 'nature' itself is not nearly so 'natural' as it seems; instead we can say that it is socially produced (Loftus 2017) (see Box 1). This does not mean that it is fictitious, but that the ways that ideas like wilderness transform places and their uses have political effects.

Preservationism wanted to preserve tracts of land intact, 'pristine', without people in it – a major influence on the spatial imagination of environmentalism today. In the late 19th century, preservationism developed in tension with early conservationism, which concerned itself with diminishing stocks for hunting and fishing, as with the potential of water reservoirs for dams in large cities. Some argue that this mode of thinking set the scene for later rearticulations of nature protection through markets, to a point where nature is even to be saved through markets (see Chapter 4). Where preservationism sought to minimise human impacts, conservationism was more comfortable with thinking through the economic value of nature and sought to manage resources for human use.[6] In this sense, two diverting paths appear in the genealogy of modern environmentalism, one tracing back to Romanticist aesthetics and the other to early sustainability science, including rational logics for planning forestry (Jonsson et al 2019). These distinct paths (spiritual imaginations of nature and economic necessity) coexist uneasily in contemporary environmentalism, pitting ideas of nature as a place apart from humans against notions of environments as resources.[7] What is notable even at the time of the making of early National Parks is that *both* aesthetic regimes help to legitimise the removal of First Nations peoples.

Wilderness aesthetics can be considered part of a broader set of spatial practices of erasure foundational to the history of European colonialism. In claiming possession over territories in the New World, European colonialists denied existing claims to those territories by native peoples, claiming land was either uninhabited or unmanaged. This is the idea of *terra nullius*, or 'land belonging to no one'. It was the legal basis for British colonisation of Australia but can also be used to describe the colonial gaze that overwrote existing systems of land management and ecological practices.[8] This is not just an accident of history, or a failure to grasp cultural differences. The rendering of uninhabited spaces as empty or wild relies on racist ideologies that cast people of colour as outside the realm of civilisation and progress (Yusoff 2018).[9] It also involves the active 'un-imagining' of communities and their histories (Nixon 2011). For it is only possible to say that inhabited land is empty by delegitimising existing forms of inhabitation, or by reclassifying them as 'surplus' to productive logics (Gidwani and Reddy 2011). Meanwhile, this reclassification makes possible a reordering of bodies for extractive purposes, exploiting categories of gender and race to render a 'docile' and malleable population (Schiebinger 2004; Bhandar 2018). Such practices are ignored in much of what happens in the name of environmentalism today, and yet they are fundamental drivers of the climate and ecological crisis, as well as spatial inequalities reproduced through environmental actions. In making this argument, we join with, and are inspired by, postcolonial geographers, postcolonial environmental humanities scholars, and anti-colonial thinkers who are acknowledged in more detail in our endnotes and in Boxes 1 and 2.

During the 17th and 18th centuries *terra nullius* forms of appropriation took place on large scales following calls from colonial administrators on both sides of the Atlantic to enclose wasteful common-land into productive plantations for European consumption. Yet such acts continue today, 'excluding people from their livelihoods' and even 'criminalising' their presence in their lived environments (Wenzel 2019: 142). This logic can also be applied not only to understand new forms of resource extraction and conservation, but to the making of new 'waste' spaces or 'sinks' that are designated as acceptable sites for the dumping of toxic materials or for experimentation with military technologies (Liboiron 2021).[10]

Connecting colonial spatial practices of erasure and dispossession with modern environmentalism is a central theme in our book. Many mainstream solutions to global environmental problems assume continued access to Indigenous lands as well as to the lands of formerly colonised countries:[11] the extraction of raw materials for 'green' technologies (Riofrancos 2019); the siting of large-scale conservation or energy projects (Dunlap 2019); the exploitation of cheap and unpaid labour, and the continued exporting of pollution, all rely on the continued existence of global peripheries and semi-peripheries.[12]

More fundamentally, we can see how particular, Eurocentric ways of framing environmental problems reproduce and intensify patterns of global inequality. In 1991, Agarwal and Narain, from the Centre for Science and Environment in India, published a powerful report on global warming and, what they called, 'environmental colonialism' (1991). Just as the architecture of global climate governance was being pieced together via new institutions such as the IPCC and UN COP summits, Agarwal and Narain identify (and challenge) a narrowly scientific accounting of greenhouse gas emissions that aggregates these chemical pollutants and assigns them to a 'shared' global atmosphere and community. Emphasising the unequal historical responsibility for these emissions, as well as the vastly different and unequal social contexts of fossil fuel combustion (for example, burning wood for cooking in India versus driving an SUV in the US), must, they argued, be the starting point for any kind of global management of climate change. Simply put: failing to place these historical and ongoing inequalities in global economic development at the foreground of climate governance ensures ongoing colonialism in the name of environmentalism. This explains the centrality of demands for (climate) debt and reparations within Global South social and environmental justice campaigns since the 1990s.

Even earlier, the world-historical movement for decolonisation that roughly begins with the end of the Second World War was making similar demands and efforts to restructure the global economy away from the social, economic and political infrastructures of colonialism. Not only does this radical political movement coincide with the birth of modern environmentalism, it can itself be understood as a struggle to decolonise ecologies that needs to be acknowledged by anyone concerned by basic questions of environmental and social justice (Ferdinand 2021). Focused in the Caribbean, the African continent and South-East Asia, these struggles gave rise to the utopian horizon of a Third World, distinct from visions of Western capitalism or Soviet communism. As we argue in Chapter 3, the practical and intellectual work of key figures like Frantz Fanon, Amílcar Cabral and Sylvia Wynter emphasises how emancipatory impulses to rethink what it meant to be human beyond European modernity, take on force when rooted in the practical reworking of relations to land and ecological relationships. This promise was short-lived as newly independent nations became the theatre for the 'Cold War' and were unable to break free of deeply inequitable (neo)colonial economic systems.

While not commonly cast as part of the history of 'environmentalism', the earlier history of anti-colonial struggles set the scene for later claims around sovereignty over resources, liberation from colonialism and capitalism, and the right to develop and prosper (Ajl 2021). The histories of such struggles become central in our book because they make demands for decolonisation inseparable from any discussions about making the planet more equitable and livable.

Given this political-aesthetic commitment, decolonial environmental politics can be understood as a collective labour to render visible and audible longer, colonial histories of ecological devastation and resistance, and to overturn prevailing Euro-American environmental imaginaries that keep privilege in place. Without this expanded perspective, we inherit a truncated view not only of history but of environmental justice. With other postcolonial and decolonial scholars, we also argue that decolonisation is about more than just formal, territorial sovereignty or state control over the use and distribution of resources. Decolonisation targets the epistemological and ontological foundations of the European colonial project, including the spatial practices of erasure and dispossession that have systematically sidelined, undermined but never destroyed, the situated and life-enabling practices of ecological repair necessary for a just and habitable planet.

Struggles for labour and livelihood

As we discuss in Chapter 4, the 1980s and 1990s was a period in which environmental concerns and debates moved to the foreground of international policy making, science and governance. The framework of 'sustainable development' promised a more optimistic future based on technological and institutional (especially market) innovations capable of growing the global economy within biophysical limits. Deforestation, overfishing, soil degradation and famine were some of the headline environmental problems that required 'solving'. In more or less explicit terms, the underlying cause of these problems within sustainable development discourse was the poverty of the majority world and their relentless, short-termist exploitation of natural resources (Sachs 1997).[13]

This familiar rehashing of Malthusian tropes of overpopulation and individual self-interest was resisted by critical scholars in the emerging fields of political ecology and critical development studies. The early work of Piers Blaikie (2016 [1985]) on soil erosion in Nepal, and Michael Watts (2013 [1983]) on famine and climate in Nigeria, demonstrated how land-use, agricultural practice, uneven patterns of environmental degradation, and social vulnerability were shaped by national and international policy making, debt and finance, globalising commodity markets, and local cultures and political systems. In a different context, Richard White (1996) wrote his provocative essay, 'Are you an environmentalist or do you work for a living?', critically reflecting on mainstream environmentalist attitudes towards work – specifically, the demonising of industrial, mechanised work, and the romanticising of small-scale, non-mechanised work.

More far-reaching and sophisticated understandings of the inseparable political economy of environmental degradation and labour exploitation had also been developed 20 years earlier. In the context of internationalist debates and efforts to de-link former colonies from world-systems in the 1960s and 1970s, decolonial thinkers were producing their own analysis that rooted the plundering of the earth and worker exploitation in (neo)colonial economic relations (Nkrumah 1966; Rodney 2018 [1972]) – such as Amílcar Cabral, as we discuss in Chapter 2. Importantly, decolonial and anti-imperialist movements, later followed by alter-globalisation movements in the 1980s and 1990s, consistently articulated the contradictory impulses between colonial and capitalist economic systems, and the flourishing of territorially based livelihoods, culture and autonomy. This has also been referred to as 'environmentalism of the poor' (Martinez-Alier 2003; Guha and Martinez-Alier 2013) or 'liberation ecology' (Peet and Watts 2004). Nor were such movements only concerned with expressing opposition to land appropriation, contamination, industrial infrastructures or proprietary biotechnologies that adversely affected them; they also modelled experimental alternatives by linking together novel farming and ecological practices with innovative modes of social organising.

Modern environmentalism has largely distanced itself from efforts to adequately account for the political-economic drivers of environmental degradation, and the situated, collective and embodied labour performed by countless people around the world as they maintain and care for their environments. Such accounts have been made forcefully by ecofeminist, Indigenous and agroecological activists and scholars, whose critiques of capital centre on the exclusion, devaluation and violent suppression of the knowledge and labour performed outside of the narrow sphere of waged labour. As the future of not just the human species, but life in general, becomes increasingly uncertain there is growing acknowledgement, following Deborah Bird Rose, that '[t]here is nothing "natural" about the continuity of life on earth, nor is continuity a process which can be taken for granted' (Rose 1996: 44). In this book, with ecofeminists like Rose, we seek to shift attention towards the persistent, multiple and enduring struggles around social reproduction that have animated many of the movements for environmental justice we describe (see Barca 2020).

Ecofeminist analysis builds on feminist critiques of the global division of labour in the 1960s, to radically recompose what is considered the 'working class' and working-class politics (Federici 2018). Labour within global capitalism is deeply gendered, racist and anti-ecological: productive work is equated to waged work, labour that is remunerated, whereas the multiple forms of care-work performed by women (for example, child-rearing), by peasant farmers in the periphery, and by nonhuman nature (for example, soil) are either unwaged, poorly paid or criminalised. And yet, as ecofeminists Veronika Bennholdt-Thomsen and Maria Mies (2000) argue so powerfully through their 'subsistence perspective', it is precisely the undervalued, life-sustaining work of women, peasants, Indigenous peoples and nonhuman nature that enables planetary life to go on. As we outline in Chapter 3, the anti-colonial, radical labour and feminist revolts of the 1960s/1970s can be considered as a rejection of the deeply inequitable capitalist organisation of work, of what counts as 'productive' and 'non-productive' work, and how this relates to ecological issues.[14]

Stefania Barca (2020) argues that the work of social reproduction is not just expressed through the care-work necessary for sustaining ecologies of land, animals, plants and human communities. It can also be extended to the defensive, collective actions taken by working-class communities when their environmental conditions are threatened or denied by industrial capitalist systems (Barca and Leonardi 2018). Environmental justice scholars have long documented how 'fence-line' communities composed of poor, racial minorities have suffered the most acute effects of contamination and industrialisation (Bullard 2018; Temper et al 2018), as well as the less well accounted for 'slow violence' of toxic legacies (Voyles 2015). Indeed, such struggles compose the four decades-long history of the global environmental justice movement (Pellow 2007). Adjacent, but often not well-articulated with the history of environmental justice, are industrial worker-led environmental politics – struggles within factories, fields and even offices where workers have been exposed to contamination and have organised to resist it (Barca and Leonardi 2018). Returning to White's argument that mainstream environmentalism has tended to demonise and dismiss industrial workplaces and workers, we can look at the connections between workplace exposure to environmental harm and broader movements to transform 'noxious' industrial capitalism (Feltrin and Sacchetto 2021). This includes the migrant fruit-pickers in the fields of California who, throughout the 1960s, objected to the widespread use of chemical pesticides because of their effects on human health (see Chapter 2). It also includes factory workers in Italy and elsewhere who, in the 1970s, not only objected to the environmental conditions of their workplaces, but to the environmental harm caused by the industries they worked in – petrochemicals, armaments and nuclear – and the possibility of reorganising industrial technology for socially useful production (see Chapter 3).

Box 2: Provincialising environmentalism

Concerns with 'who speaks' and 'who speaks for whom' have, of course, been prevalent within postcolonial studies, critical and radical geography since at least the 1980s – most notably as a result of interventions from historical movements in the Global South such as the Subaltern Studies Group. During the 1980s and 1990s the growing impact of such scholars unsettled notions of research as a means of *speaking for* those who had been marginalised by social and historical processes; an activity which, as Gayatri Spivak (1988) has famously argued, keeps repressed groups in a silenced position. In contexts of both the Global South and North, this led to fresh emphasis being placed on tracing discourses of resistance and the claims of insurgent networks (Guha 1989), together with deconstructing problematic oppositions between Indigenous knowledge and modern science (Agrawal 1995). Crucially, it centres on 'provincialising' (Chakrabarty 2009) social theories which have emerged in tandem with cultural imperialism. To provincialise social theory is to reduce its claims to universality – that is, to apply to everyone everywhere – by *situating* claims and concepts both historically and geographically, and by exploring how concepts emerging from non-dominant contexts challenge and rework their founding assumptions.

Situating claims means understanding that even concepts that have come to be regarded as 'canonical' knowledge, as relevant everywhere (like Marx's theory of capital, or Foucault's concepts of power and knowledge) were produced in particular places, and are in some sense 'local'. On the flipside, concepts produced in contexts of oppression, or non-Anglo-European sites, may speak to dynamics that are transversal – are relevant to the connections between contexts. We need to move beyond the idea of assuming that certain theories apply everywhere, whereas others can only ever be 'local'. To take situated practice seriously as a claim to knowledge is also to promote the decolonisation of scholarly discourse through attention to 'parts' that have conventionally been excluded or disregarded. In the context of everyday environmental practices this means exploring the role of *in situ* forms of knowledge – for example, that associated with the conservation of seeds and soils – in constituting alternatives, and in enabling the translation of alternatives from one site to another. As will be seen in the cases of permaculture/agroecology discussed throughout the book, the circulation of farmer-to-farmer forms of exchange are a key means through which this translation takes place.

Provincialising environmentalism therefore means questioning the ideas and concepts that get to travel everywhere, while wondering why others do not. It means noticing that the figures that get remembered in the history of environmentalism are often white and European or American, while others – as we explore in Chapter 2 – who were influential on the changing of a tide, are quickly forgotten. These ideas resonate with the work of others who have been influential in the production of this book. In particular, Guha and Martinez-Alier's (2013) book *Varieties of environmentalism: Essays north and south* explore through a variety of cases the 'rules' that mean that the poor cannot be green, that environmentalism cannot exist in the Global South – and, in contrast, the wide range of forms of innovation that go on uncounted and unremembered. In like manner, Martinez-Alier's (2003) seminal book *The environmentalism of the poor: A study of ecological conflicts and valuation* highlights the struggles of the poor or Indigenous populations involved in resource extraction conflicts around the world, to preserve their own livelihoods. While rarely associated with environmentalism *per se*, and sometimes actively disassociated from the 'white' histories of environmental innovation, Martinez-Alier claims such movements are the backbone of environmental justice.

Linking with these ideas, this book's narrative structure is based on a fundamental acknowledgement that, historically, the movements that have done the most to challenge the toxic, environmentally destructive operations of colonialism and capitalism have not described themselves as 'environmental' movements. This is not necessarily because 'environmentalism' is white, middle-class and privileged, but because the flourishing of land and people has more to do with resistance to, and escape from, oppressive and exploitative systems of colonialism and capitalism. In Chapter 4, for example, we see this demonstrated by Ken Saro-Wiwa's efforts to articulate the cause of the Ogoni people with emerging transnational discourses of environmental protection and human rights. Similarly, as we discuss in Chapter 3, movements around environmental health, the nature of work and social reproduction have manifested within the industrialised North (see Barca 2019).

Besides the more visible moments of political contestation, it is also important to observe that communities have, through all time, partaken in everyday social and material practices that involve careful relations with land, water, technology, plants, animals and materials; activities rarely classed as 'environmentalism'. The movements and practices we discuss thus illustrate how movements beyond environmentalism are often experimental and mundane, and must be counted back in as part of the broader political work of ecological reparation (see Papadopoulos 2018).

Lived ecologies and everyday experiments

Linked with the concept of *terra nullius*, which renders inhabited land empty, anthropological, postcolonial, and feminist scholarship have emphasised how the role of 'universal' science creates models for environmental action that can preclude diverse lived and embodied realities or invisibilise human relations with environments ('more-than-human geographies', after Whatmore [2006] and others). Feminist Science and Technology Studies scholar Donna Haraway ([1991] 2020) calls this blindspot a 'god-trick': the scientific idea produced through embodied observation, trial-and-error, and so on is transformed into a supposedly clean form of 'objective' truth, above the plane of subjective human error. This, Haraway tells us, is a loss, because most forms of inspired knowledge production are enriched by situated and embodied perceptions. She imagines instead an alternative ethos grounded in 'witnessing' – a form of careful, attuned involvement that does not do away with the steps of the scientific method and that, indeed, invites further investigation by others. Rather than convert environmental problems into narrowly technical problems, we need to remain cognisant of the everyday forms of expertise people have about lived-in environments, and how this alters, for example, the way that climate-induced flooding is tackled (Whatmore and Landström 2011).

There are many scientific-regulatory-policy domains where forms of scientifically mediated abstraction result in violence. Two we consider in the book are environmental conservation and environmental pollution. Both centre on what and whose bodies come to matter, and how, within projects that claim to be about the protection of all (Liboiron and Lepawsky 2022). When global environmental organisations propose solutions at scale, we see the emergence of modelling systems and programmes that are often applied by actors with little or no contextual knowledge, often through the authority of (international) non-governmental organisations (INGOs). This leads to situations where whole protected areas have been zoned before local people even know that they exist (Sundberg 1998), or communities have been fined for improper uses despite never having been informed that they are now living in a new protected area (Amador and Millner forthcoming). This, in turn, can focus attention onto attributes (for example, charismatic species – spectacular animals or plants that are easy to care about), while playing down others, and obscuring the range of options on the table.

Similarly, while harm from industrial pollution, including carbon emissions, affects everyone everywhere, it is 'unevenly universal', as Rob Nixon (2011) puts it. Yet mainstream approaches to understanding and managing pollution since the 1960s have focused on isolating specific pollutants, aggregating and rendering them as universal phenomena to be contained and regulated. Framing chemicals – whether CO_2 or dichlorodiphenyltrichloroethane

(DDT) – as discrete compounds, measured, applied and regulated uniformly, can thus replicate dominant forms of environmental conservation – obscuring different, contested and unequal relations to lively (chemical) ecologies (Liboiron and Lepawsky 2022).

Tendencies towards abstract and universal thinking have been reinforced in environmental conservation and pollution through the 1980s and 1990s as environmentalism has been annexed to the imagination and technologies of markets. Here we witness the emergence of a new 'universal', that renders value equivalent with money. This translates into an imperative to 'save' nature by making it appear in terms of dollar signs and incorporating a wealth of financial tools to cost and incentivise activities that repair damaged environments and atmospheres. This is what McAfee (1999) has aptly labelled 'selling nature to save it' (see also Arsel and Büscher 2012). Political ecologist Sian Sullivan (2013) unpacks related ideas in her article 'Banking nature', where she describes sustained efforts by conservation organisations – particularly environmental INGOs such as Conservation International, the World Wide Fund for Nature, the Nature Conservancy, the Wildlife Conservation Society and the African Wildlife Foundation – to recruit big business to conservation causes aligned with neoliberal values.[15] Of course, not all environmentalists or environmental movements think this way or believe markets are the way to solve the environmental crisis. However, many national governments, powerful international environmental organisations and international summits have invested and planned this way, such that to think otherwise is now to enact a kind of resistance – as we made clear in our opening anecdote on the 2021 COP summit. Here large-scale media performances, facilitated by new information and communication technologies, combine with proliferating mass media productions, to promote an environmental and conservation spectacle that markets the necessity of private investment in environmental conservation, while acting to engender human–nature relationships balanced in favour of financialised accumulation (Sullivan 2013).[16]

While expert knowledge is required to buttress political and economic systems that cause harm,[17] alternative forms of expertise are constantly being produced via citizen or popular science to challenge these same systems. Often beginning with affected communities marshalling their own expertise-via-experience, these forms of counter-expertise can also organise and produce their own knowledge infrastructures (Frickel et al 2010; Brown et al 2011). This idea of alternative knowledge infrastructures comes from the work of Dimitris Papadopoulos (2018), who considers collective experimentation with more-than-human natures and material processes central to the construction of modes of living beyond dominant systems of control and capitalist profit-maximisation. This is important, as the invocation is not to abandon or revile scientific methods nor the capacities of emerging technologies, but to ground experimental alternatives in a collective ethos of repair and reparation. The practices Papadopoulos investigates are compelling because they are materially innovative at the same time as they look to alter social and political horizons of the possible. One way they do this is to creatively show up what has been 'invisibilised' or rendered imperceptible in the current political situation (Papadopoulos 2018: 17). Papadopoulos, with others, calls the process that ensues 'commoning'. In the book, we read commoning as a verb for continuously remaking the world with more room for sharing, more space for alternative visions of the future and a priority on care for human and nonhuman ecologies of relation (Bresnihan 2016a).

With these productions in mind, in this book, we take care to notice how often collectives form in direct opposition to the foreclosure of voice, opportunities or livelihoods. Like Papadopoulos (2018), we emphasise the *ontological* aspects of such knowledge practices: the material infrastructures and networks of spaces, objects, technologies and people that are required for alternative meanings and subjectivities to take shape. More recently, work on Indigenous political ontologies (De la Cadena 2010) and reparative ecologies (Papadopoulos et al 2022a) makes related claims by contending that what is at stake in conflicts over resource extraction is *not* one common nature fought over by antagonistic social forces, but a multiplicity of forms of more-than-human dwelling.[18] This includes ecological movements relating to the reclaiming of land, but also alternative forms of agriculture, seed-saving and hacker cultures that are reconfiguring relations to the body, land and technology through collective infrastructures of care (Papadopoulos 2018). In Chapter 5, what we call – with other authors – 'earth politics' is a mode of collective organising that seeks to acknowledge the role of Indigenous cosmologies, lived experience and experimental technologies in future planetary ecologies.

All we want is the earth

> Some men, faint-hearted, ever seek
> Our programme to retouch,
> And will insist, whene'er they speak
> That we demand too much.
> 'Tis passing strange, yet I declare
> Such statements give me mirth,
> For our demands most moderate are,
> We only want the earth.
> James Connolly, 'Be Moderate' (2013: 84–85)

The title of our book comes from a folk song written by James Connolly, the early 20th-century Irish labour organiser, socialist and republican.

Connolly was shot, imprisoned and then executed for his role in the 1916 Rebellion against the British Empire. Though rarely taught in school as anything other than a nationalist struggle, the Irish Republican struggle against British colonial rule was also part of an international, anti-imperialist movement. Some parts of the Irish Republican, anti-colonial movement sought to construct transnational solidarities across geographic, racial, social and gender divisions – a non-sectarian political tradition that had a long history (McVeigh and Rolston 2021). This included efforts to organise across rural and urban contexts, waged and unwaged workers, feminists and cultural nationalists. As with other 'national Marxist' thinkers in colonial contexts, Connolly recognised the distinct identity and potential of diverse, peripheral working classes (Lloyd 2003). This involved working against the grain of imperial geographies – the diaspora and networks that enabled communication and organising between India and Ireland, South Africa and Scotland. The opening verse, quoted at the start of this section, captures something of this transversal and radical spirit – railing against efforts to regulate and divide what Peter Linebaugh (2019) has called elsewhere the 'revolutionary common people'. The reference to the earth roots these 'moderate demands' in the universal questions of material subsistence and need, while at the same time extending such demands globally.

Ireland secured independence from Britain in 1922, a hundred years ago. However, as many postcolonial scholars have argued, formal independence did not bring decolonisation, not for the entire island of Ireland and not in the manner envisaged and fought for by republican socialists like Connolly. The same can be said for former colonies throughout the world that secured independence in the post-war period. Yet, as we show in this book, movements and struggles for autonomy, reparations, Indigenous sovereignty and alternative futures do not let up – they move elsewhere, appear differently and continue. 'We' is a collective pronoun that is often critiqued in this book when used to efface inequalities and differences, asserting white, Western privilege in the guise of a universal category of humanity. But, as in Connolly's song, 'we' can be a powerful container for the motley masses of the dispossessed and excluded, connecting peoples across history and geography. The activists, movements and campaigns we are drawn to in this book, and our interpretation of their work, carry something of this critical generosity, connecting sites and struggles across time and space.

This book offers a critique of modern environmentalism, a movement that was instantiated through new kinds of connection, ecological awareness and solidarity between and across diversely situated individuals and collectives. As we will show, over time and through a narrowing definition of what and who counts within environmental politics, modern environmentalism has tended to foster the multiplication of managerial solutions to the exclusion of connected and vital issues of justice and power. Movements operating beyond this remit have tended to be counted outside the history of environmental politics. This, we argue, is partly what lends them such richness for the conjuncture in which we find ourselves. Through the course of the book, we notice how, within, alongside and against the grain of modern environmentalism, other movements have existed, demanding action and cultivating alternatives in the name of more just and flourishing ecological futures.

Situating the authors

The reference to James Connolly and Irish history relates to Patrick's research interests and background. For 15 years, his ethnographically informed work has brought him to diverse, 'peripheral' geographies, including fishing and farming communities in the west of Ireland and peri-urban water cooperatives in Cochabamba, Bolivia. A consistent theme in Patrick's research is how these places and the people who live in them are enrolled by and resist powerful discourses of under/development, regulatory systems and global commodity markets. His most recent work on data centres, renewable energy infrastructures and peatlands in Ireland has sought to understand the endurance of colonial relations to landscape and people within contemporary tech-driven 'green' capitalism. This is reflected in the globally unequal geographies of emerging green supply chains, where the outsourcing of resource extraction and heavy infrastructure follows existing fault lines of urban/rural, Global North/South, that have marked colonial modernity. Working with community campaigns resisting large-scale 'green' industrial projects, as well environmental justice activists and scholars in Ireland and internationally, his work aims to contribute to networks of solidarity and knowledge across spatially and historically uneven sites and struggles.

Naomi's research and contribution to this book has been shaped by her field experiences in Central and Latin America over the last ten years. Naomi is an educator, senior lecturer, researcher and creative writer based in the School of Geographical Sciences at the University of Bristol. The main themes of Naomi's research include biodiversity conservation and agroecology, with an emphasis on diverse knowledge cultures, social movements and practices of translation between sites. She is interested in the global food system, its colonial heritage(s), its embodied aspects and the ways that alternative food systems do and don't manage to help us beyond historical sticking-points. Other times, her research focuses on conservation politics – specifically the ways that 'top-down' conservation (the kind associated with big development, neoliberalism and histories of colonialism) can contribute towards dynamics of social exclusion, dispossession and even militarisation, while seeming to be a generally 'good thing'. At the same time she is very interested in the creation of experimental alternatives by small-scale collectives alongside and within contexts shaped by such dynamics. Her most recent work explores the role of changing conservation technologies, such as drones, in contributing to the militarisation of conservation spaces in conflicted environments, while also highlighting the ways that the same technologies are being used by social movements and rural communities to defend territorial rights and protect land against dispossession. She is involved in food justice through her practical interest in permaculture and agroecology and is invested in alternative pedagogies (teaching and learning practices) within and beyond the university, as part of social change practices.

We began our collaboration in spring 2011 after attending an academic conference in Nottingham, UK, on educational spaces of alterity. Finding resonance in our research interests while completing our PhDs, we began collaborating through shared reading, writing together, attending conferences and workshops, and participating together in a network of early career scholars, the Authority Research Network.[19] Prior to working on this book we have collaborated and published on themes of commons and commoning, politics of aesthetics, social movements and decolonising environmental politics (see, for example, Brigstocke et al 2021; Bresnihan and Millner 2022).

In this book we draw on our research in the field, as well as the wider historical and archival studies we have undertaken over the past ten years to situate our projects, to develop the key arguments in each chapter.

Structure of the book

The remainder of the book is structured into five chapters and four interludes, with a short coda to conclude. The chapters are roughly chronological in terms of the elements of the story of modern environmentalism they tell, although each chapter focuses on a different theme of making-visible or perceptible, and as such, relates to different aspects of the present. Thus, Chapter 2, 'Suburb, Field, Laboratory: Recomposing Geographies of Early Environmentalism', focuses on the early geographies of environmentalism in the 1960s, with an emphasis on chemicals, pesticides and the industrial food system. In this chapter we revisit the legacy of Rachel Carson, who is remembered as a seminal figure in the making of environmentalism, making clear the contributions of other, less-well remembered figures and movements, including migrant farmworkers' unions in California, and movements against the early Green Revolution in Mexico.

Chapter 3, 'Revolt Against One-Worldism: Radical Claims on Land and Work Post-1968', identifies a key conjuncture in the development of modern environmentalism. On the one hand, ideas of a single, shared global environment were emerging as the basis for international collaboration and planning frameworks. On the other hand, radical social movements taking place after 1968 into the 1970s interrupted the idea of 'one-worldism', as colonised peoples, women, people of colour, students and workers refused to be exploited within imperialist and capitalist systems. In later chapters, we show how these early concerns with decolonisation, labour and social reproduction also set the scene for more expansive routes into organising for environmental issues than modern environmentalism afforded in subsequent decades.

This is something we see in Chapter 4, 'The Right to Subsist: Transnational Commons Against the Enclosure of Environments and Environmentalism', which focuses on the consolidation of environmental governance narratives and management practices during the 1980s and 1990s – as well as the emergence of important counter-movements to protest these tendencies. We note during this period that social movements seeking to render perceptible concerns relating to resource extraction and harms relating to the industrialisation of the global food system choose to do so by names other than environmentalism. This focus also allows us to draw out the central concern with aesthetics in the book, which is to say, the ways that resonance is and is not created through social movements, collectives and the articulation of matters of concern.

In Chapter 5, 'Earth Politics: Disagreement and Emergent Indigeneity in the So-Called Anthropocene', which covers the period after 2010–2012, we contrast the way that momentum and resonance are built around 'techno-fix' type environmental solutions that place emphasis on globalised markets in environmental futures, with new kinds of imagining inspired by global gatherings of Indigenous and rural groups in the name of 'earth politics'. Invoking at once a commitment to materiality and 'earthiness' *and* to diverse cosmologies, earth politics include room for multiple possible worlds and invites dialogue between them, rather than starting from presumed agreement and universal starting points.

In Chapter 6, 'Resonance Beyond Environmentalism', which begins in the present day, we revisit our arguments by drawing out our key ideas around aesthetics, specifically focusing on resonance and disagreement. These two concepts form the building blocks for an alternative way of cultivating change to the 'environmentalist way of seeing', whose limitations we lay out in a fresh way. This brings us to our coda, which is more of an afterword.

The interludes are central to practically exploring what we mean by aesthetics and resonance. Here we show, rather than explain, key aspects of our argument, illuminating, demonstrating and story-telling through example, advertising, iconography, brands, artwork and other forms of visual narrative.

Any of the chapters in this book can be read as a stand-alone essay, with or without the notes that give more detail to the relevant academic debates. Together, the chapters detail the consolidation of modern environmentalism on the one hand, and the making of new movements *beyond* environmentalism on the other. These latter seek to establish more liveable worlds on different terms. With the interludes, which attend to the specifically aesthetic dimensions of movements and making, we hope to make vivid the limits of the modern environmental imagination, and a wealth of lively routes for organising beyond it.

2

Suburb, Field, Laboratory: Recomposing Geographies of Early Environmentalism

When Rachel Carson published her book *Silent Spring* in 1962 (Carson 2002 [1962]) to denounce the destructive effects of synthetic agricultural chemicals and pesticides on wildlife, it quickly acquired a wide Anglo-American readership, assuming cult status (Lear 1993). Retrospectively, the book's publication would be regarded as a defining moment in the making of the modern environmental movement.[1] Through the 1960s and 1970s, mass protests appealed to Carson's work as part of a rejection of the new reliance on toxic chemicals in modern industrial systems and the contamination risks synthetic chemicals posed to humans as well as bird and animal life, calling attention to the fragile interdependency of wider webs of ecological relationships. Not only the publication of *Silent Spring*, then, but its rapid and wide resonance and the subsequent scale of civic organising were a remarkable feat. Such public pressure, coupled with a liberalising of US politics in the 1960s, culminated in the passing of progressive environmental legislation in 1972, including a ban on the use of DDT – a popular pesticide. However, in this chapter we seek to demonstrate that the book's claims, however impactful, emerge from a much wider context of ideas and struggle. The environmentalism that emerged during the 1960s is inseparable from this wider context, including figures and movements that often go unremembered. Unpacking the surrounding influences helps us understand *why* Carson's work took on such resonance – but also, why other important books, ideas and figures are not remembered in the same iconic way.

The story of the public regulation of DDT is a good starting point for this book because it is one of the founding myths of modern environmentalism. It is a story seductive in its simplicity. There is a heroic figure who sounds the alarm, drawing attention to a terrible enemy invading the comfort and peace of the white, American suburb. The actions of this figure, and the mobilisation of science, results in a political response and celebrated outcome: the chemical is banned. Like all good stories, this one has a beginning, a middle and an end; a hopeful arc that leaves us secure in the knowledge that liberal political institutions are functioning well. But like all narratives, it is partial, reflecting the values and assumptions at the heart of modern environmentalism. What other figures, experiences, places, histories and chemical relations are obscured? And what are the consequences?

Importantly, because of existing racial, class and gender biases, other, contemporaneous struggles related to the same pollutants, such as protests led by migrant workers against the use of pesticides (Pulido and Peña 1998), are not as well remembered. Meanwhile, contemporaneous anti-colonial and anti-imperialist movements that also made claims about the harmful effects of US-imported chemicals and industrial forms of production are recalled in histories of anti-colonial struggles – not in histories of modern environmentalism. This is important because the resistance found in India and Mexico in relation to experimental 'Green Revolution' technologies was influential in forging agroecological and food sovereignty movements in subsequent decades – as we argue in Chapter 4. In this chapter we reconnect these entangled geographies, emphasising how they affected one another, even when they operated at a remove. In doing so, we emphasise how much is lost when these issues become separated in retrospective histories and contemporary environmental practices.

The title of the chapter evokes geographical spaces where experimentation with agricultural technologies was taking place in the run up to the 1960s (suburb, field and laboratory) and where important forms of resistance emerged. The 'suburb' is the imagined geography of *Silent Spring*, which works so powerfully on its readers partly because it implicates the homes and streets of white America. The 'field' refers to the new models of agriculture driving the introduction of pesticides on expanded scales, in the US and in nearby countries like Mexico. Meanwhile, the 'laboratory' signals the sites of science and technological experimentation with new pesticide technologies that rendered fields and bodies contact zones with unknown (or sometimes well-known) effects on bodies, land and livelihoods.

These experimental geographies, and the relations between them, helped establish aesthetics that determine which bodies are regarded as political actors for change, and which are exposed to potentially toxic effects. Revisiting the historical narratives told of early environmentalism helps us underline how these racist and colonial logics became sedimented, not only in terms of which struggles led to tangible changes, but in which struggles became *counted as environmental*. Many of the risks of the chemicals used in pesticides had been raised long before Carson's work – during the two world wars, when they were developed as insecticides for soldiers to combat malaria and everyday infestations on the frontline.[2] However, due to the separation of new domains of applied scientific practice – such as public health, chemistry, genetic modification for agriculture, and pesticide industries – this knowledge did not travel far. Chemicals like DDT took on their own hero narratives through advertising that linked their use in the wars with their capacity to wage war on the pests that plagued domestic life in the post-war era. As we trace the movement of chemicals like DDT between interlinked geographies, we also raise questions about the changing role of scientific expertise, such as: why were expert opinions about chemicals like DDT disregarded in the 1950s? How did it become possible for sidelined scientific knowledge to issue a challenge? How did disparate spheres of contestation (scientific practice, public health institutions, national legislation institutions, social movements) come together – and when did they lose their sense of association? The answers to these questions help us further understand the importance of recomposing early geographies of environmentalism to expose embedded racism and colonialism, but also to face challenges at play in the current historical moment.

To make our argument we unpack the wider context surrounding Rachel Carson's objections during the 1960s before turning to linked struggles over the use of pesticides in agricultural production in California, relating to their effects on migrant farmworkers' health. We then explore scientific debates and challenges surrounding the introduction of Green Revolution technologies, including new pesticides, into Mexican agriculture between the 1940s and the 1960s. This earlier episode helps us observe how chemicals known to be harmful were used experimentally on wildlife, 'othered' bodies (Hispanic migrant workers), and 'othered' soils outside the US. As Nixon (2011) emphasises in his account of the 'slow violence' of toxic chemicals, it has long been acceptable to expose to harm bodies and sites deemed 'disposable' by actors in the Global North, even where these harms are known to have multigenerational timelines. We argue that Carson's work was so influential partly because it was now clear that (white) *North American* lives might be at risk. While *Silent Spring* marked an important moment in collective moves to reject the use of harmful pesticides, this rejection was not itself new – parallel movements were associating these harms with long-standing issues of unequal conditions of labour and livelihood. Disconnecting these movements in the histories told about modern environmentalism sets the scene for perpetuating racist and colonial violences in the present, precisely because it insists that only white men (or sometimes women) can be actors in modern environmentalism. Wider concerns with racism and social justice were vital to the social movements driving change in the 1960s, and they are critical to questions of ecological futures and ecological justice today.

Situating *Silent Spring*

> There was once a town in the heart of America where all life seemed to be in harmony with its surroundings. The town lay in the midst of a checkerboard of prosperous farms, with fields of grain and hillsides of orchards, where white clouds of bloom drifted above the green land.
> Rachel Carson, *Silent Spring* (2002 [1962]: 5)

Rachel Carson begins *Silent Spring* with an emotive and idealised evocation of an 'average' small town in America. Carson was a powerful writer and with *Silent Spring* she sought to impress urgency and concern on her readers. Although a trained and brilliant marine biologist, rigorous in her research and arguments, she knew that feeling mattered as much as facts (Carson 2018). Imagery and narrative were part of her toolkit, and for Carson, targeting a predominantly white, middle-class audience, the image of a peaceful America threatened by an invisible, chemical threat was always going to strike a chord in an era of Cold War geopolitics and anxiety over nuclear technology.

But Carson's folding of suburban, domestic America into a new environmental awareness goes further than this. Drawing attention away from sites of wilderness and pristine beauty – those associated at that moment with conservation or preservation – and into the domestic spaces of a kitchen or backyard, she recomposes wider imaginative geographies. The undermining of Carson and her work by vested business interests and by those who refused to accept the amateur findings of a woman, has been discussed in the context of second wave feminism which developed in this period (Lear 1993). *Silent Spring* was a popular book for women's reading groups, for example, and no doubt worked to raise support for her cause among women (Rome 2003). What is less well-emphasised in these discussions is Carson's role in re-placing environment politics from the monumental, masculinist geographies of Yellowstone or the Grand Canyon, to the familiar spaces of social reproduction occupied *by* women.[3] As the nascent environmental justice movement in the US would go on to show in the 1980s, home and community were crucial sites of environmental contestation that intersected with race, class and gender (Murphy 2006).

Carson's work is also significant because it helps establish a fresh sense of the environment as a web of fragile, interconnected ecosystems that are important to protect. She also flips the Cold War narrative of threat, essential to the US imaginary of geopolitics at the time, making synthetic chemicals (rather than communism) the new invisible enemy, and modern biotechnologies designed for maximum profit, the antagonist (not Russia). It is not only the futures of wildlife that are thrown into question in this careful study and clever narrative of chemical development, but the whole question of technological futures, raising the startling possibility that they might bring about an end to human civilisation. Carson is explicit about the need to consider the ethical and political contexts within which scientific knowledge is produced and she opens fresh perspectives on the relationships between politics, society and scientific expertise, even where this meant facing unpopularity and threats from corporate interests (Oreskes 2004).

This work was seminal then in aesthetic terms: in what it made visible and how. But Carson's concern with pesticides did not emerge out of a vacuum. Objections to the industrial application of modern, synthetic chemicals had been voiced decades before the 1960s. This fact is often omitted from the popular narrative, which asserts that the regulation of DDT came about because an 'outsider' civil audience put pressure on government 'insiders' to change policy. As Russell (2001) documents, scientists were concerned about the impacts of DDT as early as 1945. Even during the First World War, concerns had been raised about its possible effects on animals after nervousness, convulsions and death were observed in laboratory cases. At the end of the Second World War, when experts were commissioned by the US government to examine its potential for widespread use in the post-war context, concerns were raised with impacts on wildlife beyond insect populations. However, due to ideological and economic interests with discovering a 'super' pesticide for industrial US agriculture, such concerns were sidelined. It was thus only when scientists and activists formed interest groups to raise concerns, and when these found growing public support, that they could exert power over the federal government, making possible the ban on DDT. This insight highlights the differences between military, industrial and scientific evaluations of pesticide technology – in this case the military articulation of DDT retained supporters for a long time and was difficult to sway (see Box 3).

It is also important to note that there were moments of civil protest prior to the publication of *Silent Spring*. For example, in 1958, 30 citizens from Long Island tried to stop the aerial spraying of DDT, bringing the issue to court using the expert testimony of biologist Robert Cushman Murphy and haematologist Malcolm Hargreaves (Russell 2001). The judges of the case found this testimony persuasive but responded with the claim that pests posed more of a threat to birds than the proposed spray programme. Carson was not the only one working to expose the spectre of ecological calamity nascent in technologies of warfare and economic development between the 1950s and 1970s, even within the Anglo-American world. She wrote to, and in alliance with, other scientists of the time, while such concerns surfaced in other works and struggles contemporaneous with Carson's biological endeavours (Gottlieb 2005). Murray Bookchin's (1974 [1962]) *Our synthetic environment*, for example, published a few months before Carson's *Silent Spring*, also describes a broad range of environmental ills associated with industrial applications of new technologies but received little attention because of its perceived political radicalism (McCord 2008).

Carson's book consequently had a very wide impact but is remembered as being more *politically* radical than it was at the time. *Silent Spring* marries a critique of DDT in biological and social terms with a literary, even nostalgic, portrayal of liberal, white America, which does not challenge its core ideological foundations. Further, though Carson challenged the outsized role of the agri-chemical industry in promoting new biotechnologies, she does not question the structural imperatives of capitalist agriculture for driving these biotechnologies. The regulation of DDT did not unduly disrupt or challenge the imperative to increase production and reduce costs to maximise profit. Instead, a new regulatory and institutional field was created, including the US Environmental Protection Agency, making the promise of scientific expertise and state oversight seem an acceptable response.

This is not to say that Carson did not want more far-reaching transformations in the agricultural system and production of industrial chemicals. But the elements of her critique that found greatest reception were those that did not require any radical challenge to the political and economic status quo. This is the risk of appealing to existing institutions and authorities. It also becomes evident when we consider related struggles at this time, also objecting to widespread use of pesticides, but not receiving the same traction – at least until wider, transversal links were established. For example, when Dolores Huerta and César Chávez, farmworkers and labour organisers in California, voiced concerns about pesticides impacting the health of Mexican agricultural workers in the US, these barely registered on the national stage (Pulido and Peña 1998). This is because the struggle was not only with dominant agri-business, but with institutionalised racism that devalued the health and lives of migrant workers. In the making and telling of modern environmentalism, the birds and bodies of middle class, white America mattered more than migrant fruit pickers in the fields.

Box 3: Pesticides in war and nature

The interest in broad-spectrum insecticides was largely created by chemical innovation in the First and Second World Wars, although pesticides have been in widespread use at least since 1868, when 'Paris green', a preparation containing arsenic, was accidentally developed (Perkins 2012). Due to shortages during the First World War, the search began for a chemical that would eliminate 'pests' across the board. While chemical pesticides fell out of favour during the 1930s, they returned to the public stage in a significant way during the Second World War, with the chemical being seen as a wonder-cure for all manner of health-related pests from 1943 onwards (Perkins 2012). After the 1940s, when the surplus production of DDT also led to an interest in its wider commercial uses, the chemical took on a 'hero' role in the battleground against lice, and later, against domestic and agricultural pests. Through parallels drawn between such wonder-cures and armies fighting against communism, concerns with its wider effects were quelled, and advertising making use of tanks and guns tackling insect fiends surged (Humphreys 1996).

Pesticides are thus embedded in long historical entanglements between technologies of war and technologies for the domestication of 'nature' – indeed, some suggest that war and the control of nature *co-evolved* through history (Russell 2001). This is an important point for rethinking modern environmentalism, because the doing of environmental science was rendered both possible *and* necessary by the demands of global military conflict. Conceptions of 'global enemies' to be fought have, for centuries, passed fluidly between military strategy and agricultural and entomology departments, while military tactics directly influenced how pests and conservation challenges would be tackled. Throughout the Cold War period a hazy distinction between military and environmental expertise persisted for such reasons: in the UK, for example, the same Oxford botanists theorising ecological precarity were those consulting on agricultural defence and informing British military commanders how to kill enemy crops in Malaya (Hamblin 2013). This exchange is remarkable, given the strong buffer between science, public policy and applied science departments that prevented emerging scientific understanding of ecologies from informing national policy before the early 1970s.

In some ways, conservation and geopolitics continued to articulate one another's agendas partly through this shared commitment to waging war on that which threatens 'life' (Biermann and Mansfield 2014). Perhaps the best-known example of this mutual influence was the widespread use of the herbicides Agent Orange and Agent White during the US war in Vietnam, produced almost exclusively by Dow Chemical. Between 1962 and 1971, nearly 19 million gallons of herbicide were sprayed in Vietnam, of which at least 11 million gallons was Agent Orange (Fallon et al 1994). Successive US administrations claimed that they were not carrying out chemical warfare as formalised by the Geneva Protocol (1925), because they were targeting plants, not people. This false separation of bodies and environment is precisely what Carson's ecological critique sought to undermine. Likewise, protests targeting the use of chemically assisted warfare in Vietnam were gradually able to surface the linkages between the chemical industry, the US imperialist wars and environmental devastation (Wenzel 2019). Revealing such entanglements is vital, given the geography of industrial testing and experimentation, which has disproportionately affected the bodies of migrant workers, the poor and those in areas of the Global South rendered 'disposable' by the rhetoric of economic development (Masco 2010; Nixon 2011). Thus, although DDT is now highly regulated and other organic compounds have replaced it in the fields, it is still used in territories beyond the US, still affecting working bodies – while other chemicals are experimentally developed in the same pattern, out of sight, or coded as 'urgent' (Agard-Jones 2014).

Industrial workers, pesticides and environmental health

> In the colour of a plum or an apricot, in the luxuriance of a bowl of grapes set out in ritual display, in a bottle of wine, the soil and sunshine of California reached millions of Americans for whom the distant place would henceforth be envisioned as a sun graced land resplendent with the goodness of the fruitful earth.
>
> Kevin Starr (1986) cited in Mitchell (1996: 134)

American historian Kevin Starr evokes a popular image of Southern California not far removed from the idealised vista of Carson's 'average' American town. Fruit, not wheat, was the primary agricultural commodity produced in this 'sun-graced' land, which Starr links to Greek and Roman agri-cultures, a 'fruit culture' that nurtures rural cooperation. What is missed in this blindingly bright vista of tidy fields and independent producers? Don Mitchell (1996), in his rich work on landscape, work and power in Southern California, paints a far more conflictual picture of the material conditions that enabled this 'fruit culture' to develop in the first half of the 20th century: the 'army of pauperized, temporary labourers that picked and canned the fruit, the monopoly conditions (rather than simple cooperation) that packing and marketing cooperatives reinforced, the dismissal of workers when the season ended' (Mitchell 1996: 20).[4] Mitchell's book documents the making of Californian agri-business from the perspective of the workers, particularly immigrant workers – Asian and Mexican. The fruit-growers in California are not quaint farmers, but capitalists who run businesses and hire workers. The land is made productive by these workers in the fields, labouring under the parching sun, cramped into makeshift camps, forced to work long hours for minimal wages.

Due to the seasonality of fruit-production, the entire enterprise was reliant on a mobile army of itinerant workers, which posed challenges. Temporary camps were not safe or secure, there was minimal hygiene. It was around these issues that militant organisers fanned discontent among workers drawn from all over the country and beyond in the early 1900s. As early as 1913, the Wheatland Riot involved a violent confrontation between striking agricultural workers and fruit growers. The strike was organised to protest poor working conditions not just in the fields but in the camps (or lack thereof), where the workers slept, ate and lived between picking. The response by the Californian authorities was to devise blueprints and budgets for 'model' labour camps designed to prevent 'unrest'. As Mitchell argues, this technical approach to managing worker grievances, expressed through new discourses and practices of environmental and occupational health, would characterise state responses throughout the 20th century. Mitchell's book stops in the 1930s but his thesis can be extended to consider the emergence, and response to, new worker concerns centred on the harm caused by toxic pesticides.

Widespread pesticide use in California from the 1940s was not just an effect of wartime chemistry. Californian agriculture was shaped by the pursuit of profit. This was reflected in social relationships and institutions, such as private land ownership and exploitative labour relations, but also in the changing ecology and landscape (Nash 2007). Farmers had introduced new plant varieties selected and grown to maximise returns. One consequence of the introduction of unsuitable crops to the location was the introduction of new insects. 'By the early twentieth century', historian Linda Nash writes, 'California fruit growers were contending with scales, beetles, thrips, moths, phylloxera, aphids, spider mites, and peach-tree borers' (2004: 206). The availability and promotion of new organophosphate compounds to the domestic market after the war met farmer demand that had been growing for decades. By 1963, a year in which many more Americans were now familiar with DDT, and over 16,000 different pesticides had been registered in California (Nash 2004), farmers were spraying their fields with multiple chemicals, several times a year.

It was not long before agricultural workers, exposed to cocktails of often toxic chemicals, began reporting incidents of acute ill-health: sore eyes, skin conditions, swollen mouths, extending to fatalities as early as the 1940s. Yet, as in the DDT case, there was little attention paid to the problem until the 1960s, when links between chemical pesticides and disease were made perceptible beyond individual experiences and concerns through the organising efforts of fruit-pickers.

In the same year that Carson's book made the connections between DDT and the environment visible to a US public, Dolores Huerta, born and raised in the US, and César Chávez, a Mexican migrant to California, co-founded the National Farm Workers Association, which would later become the United Farm Workers (UFW). Huerta and Chávez began organising in the early 1960s in the southern San Joaquin Valley. The pesticide problem surfaced early on for labour organisers, and the UFW set up a makeshift clinic during the first major strike (Pulido and Peña 1998). One of the challenges, however, was that, while the workers, union and certain healthcare bodies knew that pesticides were causing sickness among workers, the extent of these health risks and problems were not apparent. Nor was it clear how to collect the evidence necessary to communicate with public health authorities and build wider social support.[5]

In the 1950s, there was limited research or awareness about the effects of pesticides on human health (Nash 2004). The dominant scientific understanding was that chemicals could not enter the body, and thus the only way farmworkers could be affected was if they ingested them. One minor subfield where connections between the environment and human health *were* being made was occupational health, and it was here that concerns were raised about the ill-effects of organophosphate compounds on humans after the Second World War (Nash 2007). For migrant workers, though, the main problem wasn't the lack of scientific expertise to help make sense of their situation, it was structural racism. Operating in more or less explicit ways, this racism ultimately regarded the migrant population as *inferior*, manifesting in biological (weak and more susceptible to sickness) and cultural (less educated in basic hygiene) discourses that explained away their illnesses and suffering.

A further challenge facing sympathetic health professionals attempting to gather evicence was the constant mobility of agricultural work. Given that workers moved freely between different fields and farms, using different chemicals, in different quantities, establishing clear correlations between exposure and ill-health was basically impossible. As poisonings persisted and accurate monitoring and data continued to prove elusive, the response from public health agencies was to call for more research, 'as if one more field study would finally allow them to describe and control how chemicals moved from a complex environment into an individual body' (Nash 2004: 218). Public health officials were thus able to dismiss concerns through claims that migrant populations and chemicals were challenging to monitor (Nash 2007). This is the 'problem of uncertainty' that has fundamentally shaped the politics of environmental health and expertise from the 1950s to the present (Murphy 2006). By demanding 'clear' evidence, and isolating particular agents and their effects from wider social contexts, action can continuously be deferred and delayed (Boudia and Jas 2014) (see Box 4).

The farmworkers of Southern California were consequently faced with the limitations of official science and regulation in a context of uncertain pathways of chemical pesticides and their uncertain effects on human health. The UFW also recognised that the determining factors behind pesticide use were not (a lack of) scientific expertise and awareness, but political and economic systems, namely, the unequal power of Mexican workers vis-à-vis farmers and the Californian state. For these reasons the UFW created its own Health and Safety Committee to advise the union on its strategy around pesticide use and how workers could protect themselves through contractual obligations placed on the farmer (Pulido and Peña 1998). This work informed lobbying and campaigns such as gaining public access to pesticide reports; banning the use of specific pesticides; and transferring pesticide regulation from the California Department of Food and Agriculture to the Department of Public Health (Pulido 1996).

By 1969, the UFW was able to bring several farmworkers to testify before the US Senate. They spoke about the extent of spraying, chronic headaches and skin conditions, nosebleeds and seizures (Nash 2004). Understanding the limits of their testimony within wider hierarchies of expertise, the UFW also presented certified data collected by the medical profession on the abnormally high levels of organophosphate pesticides in children of Tulane County. Nash (2007), who we draw on extensively here, captures the inventive work of citizen and worker-scientists in this case, who worked painstakingly to collect and repurpose expert knowledge to their own ends: the presence of chemicals in the bodies of white children demonstrated the porosity of the human body, and thus the dangers posed to migrant farmworkers through their everyday exposure.

When this organising labour coincided with wider cultural receptivity to chemical risks and the environment, in part brought about by Carson, one of the UFW's most innovative and significant campaigns became possible: a consumer boycott of Californian grapes. The boycott required a shift in how such fruits were perceived by the consuming public. Against the ideological bulwark of a kind of idealised 'fruit culture', the UFW leveraged existing fears and uncertainties about chemical pesticides to resignify these fruits as deadly. If the sickness and death of Mexican workers didn't matter to the US public, then the health and well-being of the people who consumed the fruits of their labour and bodies might. This appeal was similar to that made by Carson; an appeal to self-preservation and security, but also to the norms and attitudes of white, middle-class people who would object to chemical impurities in their food.

The largely powerless UFW did not organise the grape boycott to 'raise awareness' among middle-class consumers but to question the reputation of growers and disrupt their commercial operations. This was a tactic aiming to force growers to the table to negotiate contracts with workers, including over the use of pesticides (Nash 2004). Pesticides thus accrued importance in that they became a means to connect with a wider set of interests and constituencies. Meanwhile, the boycott functioned to invoke wider, *environmental* concerns over pesticides – even though the pesticides that were of most concern to consumers (namely, DDT), were not the ones that posed the greatest danger to farmworkers. The UFW emphasised the uncertainties and risks that these chemicals posed to everyone, worker or consumer, and were thus able to achieve their own priorities via the interests of others.

In 1970, California farmworkers won the right to unionise and their first collective bargaining agreement included an unprecedented Health and Safety Clause. This was undoubtedly a huge achievement for a worker organisation that had been formed for less than a decade and which faced powerful opponents in the organised growers of California (Pulido 1996). The Health and Safety Clause included significant measures such as the ban of 'economic poisons' (including DDT), and the availability of all information on pesticide use to Health and Safety Committees made up of workers. Organophosphates were not banned however, they were regulated *just enough* to ensure business as usual could continue. The chemical industry and agri-business continued to evade the slow pace of science and regulation, and migrant agricultural workers continued to get sick and die as a result.[6] The efforts of the UFW, and figures like Carson, were important but the structures of racial capitalism halted any real advance in environmental justice.

Box 4: Environmental racism and Black ecology

In the US, environmental justice and environmental racism are synonymous. The history of the environmental justice movement is often traced to the campaign by an African-American community in Warren County, North Carolina in the early 1980s (see Bullard 2018). Objecting to the siting of an Environmental Protection Agency-approved waste dump for soils contaminated with toxic PCBs (polychlorinated biphenyls), the community resisted the decision on the grounds that it was a risk to their health and a clear instance of 'environmental racism', a term coined by the organisers. Connecting with the Civil Rights Movement in the Southern US, the conflict in Warren County made clear that questions of environmental health, including the expertise that guides planning, testing and regulation of toxins, was inseparable from social inequalities and structural racism.

As the organising of the United Farm Workers in the 1960s demonstrates, links between race, health and environmental protection were already being made 20 years earlier. And from the outset, the environmental movement in the US faced criticism for its perceived 'whiteness'. In the lead up to the first Earth Day in 1970, for example, Black politicians were quick to criticise the environmental movement as having nothing to do with the urban issues facing Black communities (Ferguson 1970). While undeniably true, there were others who sought to redefine environmental issues in Black terms.

Freddie Mae Brown was a community worker in St Louis and chair of the Metropolitan Black Survival Committee. She argued that the 'definition of environmental pollution must include areas that white people largely ignored. Issues such as the non-existence of adequate housing, lead poisoning, health, unemployment, and other problems that make up the general slum habitat' (quoted in Ferguson 1970). In 1970, for Earth Day, Brown wrote a series of skits with the assistance of Wilbur Thomas, a public health scientist at St Louis's Washington University. Entitled 'Black Survival', the text presents a series of vignettes where Black families face health problems caused by rat infestations, air pollution and lead paint poisoning (Gioielli 2019). Importantly, by the end of the scenes, the young protagonists have made clear links between the environmental problems they face and the inequalities of the post-war American city.

In a short but powerful essay written in 1970 entitled 'Black ecology', sociologist Nathan Hare also connects exposure to air pollution with the racialised spatial segregation in US cities. As with Brown, he emphasises the problem of inner-city traffic and proximity to newly built expressways as a specifically Black working-class environmental problem. Significantly, he cites the work of Fanon, concluding with the line: 'The real solution to the environmental crisis is the decolonisation of the black race' (Hare 1970: 8).

Decolonisation and environmental justice in the US context can be traced back to waves of resistance to settler-colonialism by Indigenous peoples. In the 1960s and 1970s, Red Power and Pan-Indianism grew in the US, with Vine Deloria Jr. its most well-known spokesperson. Echoing Nathan Hare's essay and writing in the same year, Vine Deloria Jr. also argued that the solution to the problems (including environmental) facing North America was decolonisation: 'It just seems to a lot of Indians that this continent was a lot better off when we were running it', he wrote in the *New York Times* (cited in Red Nation 2021: 33). Like the Black Power Movement, Red Power demanded greater civil rights for Indian peoples, including access to resources, education, housing and healthcare, as well as greater self-determination through protection of territorial rights (Karuka 2017) – the occupation of Alcatraz and Wounded Knee were emblematic of this demand for the return of Indigenous land and a reclaiming of Indian culture, legacies that have returned with such force today (see Chapter 5). The intersection of racial and environmental justice has continued to be central to movement organising in the US, including LatinX (De Chiro 1996) and Indigenous communities (Whyte 2016).

Mexico as a field-lab for Green Revolution technologies

In parallel with US-based struggles over whose bodies could be exposed to which harms, and what limits should be applied to the use of chemicals for the elimination of 'internal enemies', the same chemicals were being applied to fields in the Third World with much less fanfare. From the 1940s onwards, Mexico was used as a kind of test-laboratory for ideas about modern agriculture, including genetic modification, fertiliser technologies and chemical pesticides. The intentions behind this intervention were framed – in the language of US development institutions – as compassionate: when Vice-Chancellor Henry Wallace visited Mexico he expressed shock at the hours of manual labour it took Mexican farmers to produce a bushel of corn, in contrast with large farms in Iowa he was familiar with (Marglin 1996). This visit formed the basis for the conceptualisation of the Mexican Agricultural Program (MAP), which later became the blueprint for modernising agriculture across Latin America (Edelman 1980). However, in practice these techniques were pushed aggressively: Mexico and other Latin American governments were told that if they did not accept a 'green' revolution, they would see a 'red' (which is to say, a communist) one (Bebbington and Carney 1990). Despite the warnings raised by key social scientific and scientific experts, the subsequent transformations inflicted significant damages on agrarian communities, soils and genetic seed variety across Latin America and beyond.

In this section we follow the way that objections were raised as the MAP took shape. These objections appear to be distanced from early environmental struggles in the US. Nevertheless, we show how the connections and associations forged at this time gave important shape to new collectives emerging in the 1960s, under banners such as agroecology. We are not saying that such movements directly influenced early environmentalism as it was emerging, however, we do contend that, first, wider movements were making related claims about agricultural chemicals, which influenced the context figures like Carson were working in, and, second, that such movements connect issues of livelihoods, labour and critiques of (neo)colonialism with the environment – issues that become progressively sidelined as modern environmentalism develops.

Green Revolution technologies sought to intensify agricultural production around the world as a response to growing levels of global hunger, focusing on the innovation of genetically modified seeds that were high-yielding and disease resistant. Coordinated by scientists such as US agronomist Norman Borlaug through the 1940s and 1950s, these technologies involved the development of high-yielding varieties of cereal grain alongside the expansion of irrigation infrastructure; the consolidation of land holdings; and the distribution of hybridised seeds, synthetic fertilisers and pesticides to farmers. This led to a 'boom' in new chemical fertilisers and pesticides to stabilise mass crops like hybrid corn, which could not reproduce itself year on year like traditional crops, and drained the soil of nutrients. Mexico is often seen as the birthplace for this 'revolution': between 1940 and 1965, Mexico's production increased fourfold, while the country moved beyond relying on imports for its food needs (Sonnenfeld 1992). It was subsequently heralded as a model by the Rockefeller Foundation, the US Agency for International Development, the World Bank and other agencies, to promote Green Revolution technology packages around the world (Sonnenfeld 1992).

While leading to dramatic increases in crop yields, the deployment of these new seeds on huge scales immediately impacted rural economies and ecologies. The new benefits did not accrue evenly across the Mexican population, tending to benefit larger scale farmers with significant investment capital, and urban industrial capitalists. The emphasis on single seed varieties quickly narrowed the diversity of genetic varieties under cultivation, while less commercially orientated farmers were forced to compete with new models of mechanised high-output agriculture. Meanwhile, the new varieties required higher and higher inputs of pesticides, as monocropping made entire harvests more vulnerable to infestations and chemical fertilisers (as the new varieties drained the mineral content of soils more rapidly) (Patel 2013).

The MAP transformed the economies and biodiversity of rural areas in Mexico, focusing production on a handful of adapted corn or bean varieties from thousands of cultivars, the seeds of which had to be bought (Kloppenburg 2005). By 1960, 15 million hectares of *ejido* (communally-managed) land had been abandoned (Sonnenfeld 1992). Tens of thousands of people were displaced, including many Indigenous people using long-standing, sustainable agricultural techniques. This gave rise to a new role for Mexican workers in American fields, alongside the new role of fertilisers to sustain the new levels of production, connecting the three domains discussed in this chapter around their hidden 'externalities' – which is to say, the way that new levels of industrial productivity held costs that were shifted from one set of bodies or environments to another. Thus, for example, when the US banned the imports of fruits and vegetables sprayed with DDT, agriculturalists in Mexico began using other pesticides which have shorter half-lives (and thus allow the produce to pass border inspections), but which are more acutely toxic for agricultural workers and their families who come into immediate contact with them. Research continues to document numerous deaths in Mexico attributed to pesticides, most notably among Indigenous women and children (Lopez-Carmen et al 2022).

Importantly, the Green Revolution also transformed how global problems were tackled. Borlaug was invited to India by the adviser to the Indian minister of agriculture in 1961, when the country was on the brink of mass famine, but also to Mexico in the 1950s, after the visit of the vice-president of the US, Henry Wallace. Such exchanges, critically, consolidated a solutions-focused 'white science' that functioned to equate whiteness with scientific superiority and indigeneity with underdevelopment (Eddens 2019). Over time, the real winners of these new biotechnologies were large-scale fertiliser, pesticide and seed companies based in the US, whose economic agency also increased the geopolitical power of the US government in Latin America and beyond (Kloppenburg 2005). Through the MAP, later used as a model for countries throughout Latin America, Mexico became a wheat exporter, while commercial farmers and US-based fertiliser and pesticide companies extracted profits that allowed them to become powerful actors in Latin America (Cullather 2004). Yet alongside this history, building on dissent voiced as early as the 1940s, networks of agronomists, peasant farmers, geographers and anthropologists worked to document and protect the infrastructures of traditional farming now seen to be under threat. Since the 1960s, these networks have become associated under the term agroecology, which denotes approaches to agriculture that value ecological factors over other metrics of success. Agroecological networks have long prioritised the experience and expertise of peasant farmers, and focus on devising principles for *in situ* experimentation to evaluate the usefulness of traditional and innovative farming techniques.[7]

The model of knowledge production at the core of these networks has been around far longer than the Green Revolutions they seek to overturn, however. Astier et al (2017) note that *in situ* experimental solutions have developed continuously throughout history, and were consolidated in important ways during the 1940s, when prototype ideas for Green Revolution technologies were still under debate. American geographer Carl Sauer was one important white critical voice: when consulted by the Rockefeller Foundation on the US plans for Mexico, he rejected them outright, anticipating significant damage to local ecologies and economies (Bebbington and Carney 1990; Harwood 2009; Astier et al 2017). Sauer (1941, cited in Bebbington and Carney 1990: 35) advised strongly against the plan, warning of the dangers of applying agricultural science 'to recreate the history of US commercial agriculture in Mexico' and advocated a more 'bottom-up' process that respected the integrity and complexity of already-existing agricultural economies and biodiversity systems. Sauer recommended instead that the improvement of the genetic base of agricultural crops be predicated on an understanding of the relation of such work to the poorer segments of the society (Oasa and Jennings 1982: 34), for example, by considering the role of various foods in their kitchens. The nutritional practices of the Mexican peasantry, or *campesinos*, were, he claimed, excellent as far as their incomes allowed. The same was said to be true of their agricultural practices. Sauer's objections were ignored, however, and the MAP became an early prototype of Green Revolution technologies.

Sauer's public comments also chimed with concerns voiced by agronomists in Mexico such as Edmundo Taboada and Edmundo Limón, who emphasised that the new hybrid varieties interrupted traditional practices of seed-saving (the new varieties had to be bought anew each year and could not self-reproduce). Together these critical voices called for alternatives that were not so reliant on external inputs and would protect important parts of Mexico's *campesino* economy (Astier et al 2017). In the decades that followed, wider webs of activists and scholars based in affected regions would connect the ravaging of land with longer histories of colonialism and give rise to new kinds of agroecological training schools and networks (Rosset and Martínez-Torres, 2012; Holt-Giménez and Altieri 2013) as well as postcolonial activism (Gupta 1998) (we return to this theme in Chapters 3 and 4).[8]

Taboada was a particularly important figure in these earlier decades of critique. Like Amílcar Cabral in Guinea (see Chapter 3), Taboada was dedicated to experimental agronomy in field stations where long-standing practice was seen as the basis for future science, rather than quick-fix industrial solutions. This is evident from his claim that 'useful science, manageable, operable science, must grow out of the local laboratories of … small farmers, ejidatarios, and indigenous communities' (Taboada, cited in Hewitt 1976: 19). Before the arrival of US-based development institutions, movements led by agronomists like Taboada had linked national agriculture strategies with the active participation of the peasantry and a valuation of their expertise (Oasa and Jennings 1982: 36). In 1947, an Experimental Stations Office was converted into the Instituto de Investigaciones Agricolas (Institute of Agricultural Research), which continued to function until 1960. However, when the MAP arrived, its development institutions funded new initiatives that valued global expertise: most notably, the lavishly endowed, US-influenced Rockefeller-supported Office of Special Studies (OSS), became the centre of MAP and Green Revolution programmes. Where the peasant-focused Instituto de Investigaciones Agricolas concentrated on developing improved open-pollinated maize varieties for poor farmers (Oasa and Jennings 1982) through rural conversations, the OSS emphasised the need to raise yields through the use of hybrids grown with mechanical irrigation. Due to its funding and international interests in its success, the OSS gradually overshadowed its competitor, and, after 1960, agronomic research for smallholders in dryland areas received relatively little attention (Edelman 1980).

As we discussed in the introduction to this book, such disagreements are highly revealing about whose knowledge gets to count in the making of (in this case agricultural) institutions and practices. The alternative model of agricultural science in the Mexican context was sidelined over time by the American model, fostering a storyline that affords this second model more agency in making modern environmentalism than it perhaps is due. The Mexican paradigm, however, still offers ingredients for what we have called a disagreement over the givens of the situation – in other words, it creates the basis for alternatives that are based in different ways of understanding and narrating the problem.

The 1960s, sometimes considered the moment of emergence of new models and participatory solutions, is thus better understood as a second wave of opposition to the agro-industrial model in Mexico and in other experimental 'field-laboratories' of the Green Revolution. Seminal works from this period record a new scale of concerns over the foreseeable consequences of the agro-industrial model and demonstrate an explicit recognition of the value of webs of local botanical, entomological and agricultural knowledge. Key authors, including Efraím Hernández Xolocotzi (1913–1991), Arturo Gómez Pompa (1934–), José Sarukhan Kermez (1940–), Miguel Ángel Martínez Alfaro (1942–2007) and Rafael Ortega Packza (1944–) attempted to understand and record this knowledge in its own terms for these reasons, inspiring subsequent agroecology seminars and early university programmes (Gliessman 2013; Astier et al 2017).

In common narrations of early environmentalism, it is often suggested that agroecology emerged *after* the 1960s, in the wake of calls from figures like Rachel Carson. However, this parallel geography suggests that the seeds of agroecological movements came together as part of critiques made as Green Revolution technologies were being first introduced – decades earlier, and in Global South contexts. In other words, an important thread of what we now call environmentalism was being woven at the same time as objections were being translated into public spheres in the US by Carson and her colleagues. Although these Mexican histories have been sidelined in accounts of modern environmentalism, they raise important questions for environmental movements around wider geographies of knowledge and expertise: whose objections get to count in the making of agricultural futures? Who and what is allowed to count as part of environmentalism, and why are some actors and actions rendered outside its histories? Distinctively, the domain of proto-environmental practice we trace here also keeps questions of human labour and livelihood central, at the same time as raising issues with the limits and damages of modern-industrial technologies, such as those employed in new forms of capitalist agriculture.

Conclusion

Rachel Carson did not deliberately exclude the workers of Southern California or the agricultural transformations happening in Mexico from her environmental fable of modern North America, and that makes it all the more significant. The popularity of her book was, in part, because it was consistent with an environmental sensibility that pivoted on the idea that nature and culture were separate domains. The early environmentalism that grew up around these ideas fought to defend a nonhuman environment that was seen to be at risk from new kinds of human exploitation, yet was imagined to be distinct from entanglements with race, class, patriarchy, and (neo)colonialism. Over time, this narrative has fostered a storyline *about* environmentalism that makes books like *Silent Spring* seem like the first articulations of an idea about the limit to the new chemicals being transferred from war science to agriculture and public health. In fact, as we have shown, objections were being raised in Mexico, India and other zones where Green Revolution technologies were experimentally introduced, as early as the 1940s. Scientists also raised concerns with DDT when it was first introduced, and agricultural workers' movements were among the first to highlight its toxic effects on human bodies.

What is necessary, we argue, in any revisiting of early environmentalism that aspires to move beyond the blindspots of modern environmentalism, is to notice which actors and movements are remembered as the 'heroes', and which are sidelined over time. Moreover, we can observe how this selective remembering process mirrors and reinforces a dynamic that renders particular bodies and fields as zones of experimentation, and others as 'at risk' and to be preserved. Some bodies, more than others, were exposed to toxic chemicals through the modernisation of agricultural technologies, and some places, more than others, were used as sites of experimentation for technologies of war, industrialised food production and public health (see Box 3). As we note the common points of resistance between these sites, we also want to observe the way that retrospective re-memberings render only certain (usually white, Anglo-American or European) figures as authors and authoritative in the making of modern environmentalism. However, the same logic was at play in rendering certain sites and bodies as appropriate test-sites for new industrial and military technologies at the time. While the Pacific was deemed a legitimate testing-ground for nuclear weapons during the post-war period (DeLoughrey 2011) India and Mexico became experimental sites for testing novel agricultural technologies and techniques through what became known as the 'Green Revolution'.

Thinking this way asks us to look again at what we call 'the environment'. Historically, environmentalism and agriculture have tended to be thought about using entirely separate paradigms (Agrawal and Sivaramakrishnan 2000), with the former framing Global North-led international organisations as experts over tracts of 'pristine' forest or land to the exclusion of those who live there, and the latter empowering industrial models for enhanced production among the supposedly ignorant poor. Both have systematically sidelined *campesinos*, rural people, Indigenous groups and women, and both have worked to make issues of human livelihood appear separate from questions of environmental degradation. Yet, as the work of anti-colonial thinkers such as Frantz Fanon, Aimé Césaire, Suzanne Césaire, and Amílcar Cabral remind us, matters of soil and interspecies health are crucially connected to the dignity and labour struggles of oppressed peoples (DeLoughrey and Handley 2011). In the next chapter we will explore how these related struggles, alongside femininst and worker-led movements in the Global North, further expand our sense of what has, and could be, considered environmentalism.

First Interlude: Green and White Dreams

A blank canvas and a green rugged field
Lined with a white picket fence,
Only a little ground to clear to make it ready,
This is the American Earth!
The great hand of Union Carbide
Erects a great wall, a dividing-line
Between the past and the rocket-fuelled future
The white worker and his white wife
Stand by: We are on our way to freedom
They say, pointing ahead, where
Two roads diverge – one a highway,
The other, a low road.
Dupont and his cellophane baby
Wave beside Shell-powered engines
Blessing the way to the countryside:
Ours. Just don't look out the window, kids!
Those people with placards,
Those people falling in the fields –
They are not with us.

★★★

In this Interlude, we juxtapose images that circulated around the production of plastics, petroleum and chemicals during the 1950s and 1960s, in connection with new dreams of progress. With a focus on companies and chemicals linked with the UK and US, we have chosen images that connect histories and geographies of war, imperialism, and race, with ideologies of progress, hygiene and the countryside. Alongside these visual stories we place records of protests that erupted – often decades later, when toxic effects on bodies and landscapes had become clearer – in relation to hidden aspects of these same images. The toxic effects of waste incineration in the making of plastics; the effect of Agent Orange – the chemical used in new weapons during the Vietnam War – in the soil on bodies; the effects of DDT on ecologies; the effects of war on communities: new forms of aesthetics emerge to denounce such effects and render them perceptible to a global audience.

Figure 1: Du Pont Cellophane advert, 1954

Source: E.I. du Pont de Nemours & Company, I_43_32, Advertising tearsheets in Advertising Department records (Accession 1803), Hagley Museum and Library, Wilmington, DE 19807. Courtesy of the Hagley Museum and Library.

DuPont de Nemours, Inc., commonly known as DuPont, was founded as a gunpowder manufacturer in 1802 by E.I. Du Pont. Before merging with Dow Chemical in 2017, DuPont was the world's largest chemical company in terms of sales.

Cellophane was invented in 1912 but became widespread after DuPont developed a moisture-proof version in 1927. In the 1950s, DuPont grew their business through the mass production of Cellophane and another well-known synthetic material: Teflon. Both products are built out of a group of at least 4,700 synthetic chemicals called PFAs, short for perfluoroalkyl and polyfluoroalkyl substances, which make surfaces resist stains, water and grease. PFAs are also toxic, linked to an increased risk of cancer, endocrine disruption and weakening of the immune system. They are often referred to as 'forever chemicals' as they remain in bodies and environments, accumulating over time. For decades, DuPont dumped PFAs into the Ohio River in West Virginia, killing farm animals and poisoning the water of surrounding communities (see Altman 2019).

Figure 2: 34th anniversary of the Bhopal tragedy, 3 December 2018

Source: Rohit Jain, Bhopal Medical Appeal

Every year, protest rallies and all-religion prayer meetings mark the Bhopal gas tragedy that killed over 5,000 people after the leak of 40 tonnes of deadly methyl isocyanate gas at Union Carbide's pesticide plant on 2–3 December 1984 (see Figure 2). Though unofficial estimates claimed that the death toll due to the tragedy had crossed 25,000, official figures stood at 5,295 for whom the government had compensated. Many of these deaths were 'delayed deaths', caused by gas poisoning, with tens of thousands of survivors also struggling with cancer, blindness, respiratory difficulties and immune and neurological disorders. Protests take place in the old Bhopal area where the defunct Union Carbide factory is situated, demanding justice and adequate compensation was demanded for tens of thousands of suffering survivors of the world's worst industrial disaster.

FIRST INTERLUDE

Figure 3: Advertisement for chemicals produced by Union Carbide, c. 1954

Source: Union Carbide/Alamy

The Union Carbide Corporation referenced in Figure 3 is a US chemical corporation which began life in 1917 with the production of ethylene from natural gas liquids. It was a key instigator of the modern petrochemical industry, developing military and commercial applications. The company is linked with two of the worst industrial disasters of the 20th century: the Hawk's Nest Tunnel disaster in West Virginia, in the 1930s, memorialised in Muriel Rukeyser's poem *The Book of the Dead* (1938), and the Bhopal disaster in 1984.

Figure 4: 'DDT is good for me-e-e!', *Time Magazine*, 30 July 1947

Source: Phil Allegretti Pesticide Collection (Box 1), Science History Institute, Philadelphia

The text in Figure 4 assures the US public that DDT promotes, rather than harms, health. This was part of a sustained public campaign, which later included staged demonstrations of DDT spraying – intended to reduce mosquito populations causing malaria and other diseases – on school children eating their lunches, in swimming pools and in other public places.

Figure 5: Flit advertisement, 1930–1940, Dr. Seuss Advertising Artwork

Source: Special Collection and Archives, University of California, San Diego Library

As detailed in Chapter 2, the damaging effects of DDT on human health and wildlife were known by some experts at this time. One of the major manufacturers of DDT was DuPont (see Figure 1). Along with other chemical companies, DuPont undertook promotional campaigns for pesticide use and sought to undermine Rachel Carson's credibility and the public impact of *Silent Spring*.

Before Theodor Geisel, who later became known as Dr Seuss, found fame as a beloved children's book author, he landed his first steady illustration job as a cartoonist for *Judge*. During this time, he drew a cartoon depicting a medieval knight in bed, facing a dragon who entered his room, lamenting: 'Darn it all, another dragon. And just after I'd sprayed the whole castle with Flit'. According to an anecdote in Judith and Neil Morgan's book *Dr. Seuss and Mr. Geisel* (Morgan and Morgan 2009), the wife of the advertising executive who handled the Standard Oil Company's account (which owned Flit) saw the cartoon, and urged her husband to hire the artist. This was the start of a popular 17-year advertising campaign where 'Quick, Henry, the Flit!' became a popular catchphrase. Images drew on military tropes circulating during and after the Second World War, making connections with the war on pesticides and the war against the Axis powers. DDT was also used as a pesticide in the trenches.

As farmer and author Will Allen (2008) notes in *The War on Bugs*, Seuss helped America become friendly with the poison of pesticides (including DDT) and grow comfortable with the myth that pesticides were necessary to live well. As Allen notes, many believe the positive, pro-environmental themes of Seuss's children's books actually stemmed from Geisel's desire to clean up his legacy of helping to introduce chemicals into the food supply.

Figure 6: Protests against crop dusting, date unknown

Source: Crop dusting photograph, Sarah/Flickr, Creative Commons licence CC BY 2.0

Although DDT was banned in the US in 1972, protests began much earlier. Rachel Carson was one of those who pointed out its effects on wildlife and people after receiving letters from neighbours of bird sanctuaries who had noticed the birds dying.

Figure 7: British advertisement for Shell petrol, featuring wildlife in Wales painted by Maurice Wilson, 1955

Source: Shell/Alamy

In 1954, the Shell Oil Company released an illustrated calendar series entitled 'Key to the Countryside'. The calendar was part of a wider promotional and advertising company that depicted the British countryside as an untouched, rural idyll. At this time, the automobile industry was expanding across both sides of the Atlantic, sold on the promise of individual freedom; petrol and private cars were the key to the countryside.

After the Second World War, Shell also launched new exploration programmes in Africa and South America in a bid to keep up with shifts in energy production and consumption, including the demand for petrochemicals from the oil refining process (used for mass-produced pesticides and fertilisers, for example). Oil deposits were found in Mexico, Borneo and Nigeria. In 1956, Shell discovered the first commercial oil field at Oloibiri in the Niger Delta and started oil exports in 1958. Two years later Nigeria declared independence from British rule.

Figure 8: Greenpeace activists protest at a Shell gasoline station in San Francisco, California, 1995

Source: Melanie Kemper/Greenpeace

Over a 50-year period, Shell has extracted millions of barrels of crude oil from the Niger Delta and billions of dollars. At the same time, most of the Niger Delta population has not benefited, and environmental devastation has been intensive and extensive. The Indigenous Ogoni people have been particularly badly affected – polluted soils, waters and air, caused by oil spills, leakages and flaring have made *their* 'countryside' unliveable. In 1990, the Movement for the Survival of the Ogoni People (MOSOP) began campaigning for an end to the environmental devastation, greater political rights and reparations from Shell (see Chapter 4).

Figure 9: Ogoni Solidarity Ireland, County Mayo, 1995

Source: photo © Ken Saro-Wiwa Archive Maynooth University Library (PP/7/107)

In the early 1990s, Ogoni Solidarity campaigns began around the world through the effective networking and activism of MOSOP and Ken Saro-Wiwa. An Irish nun, Sister Majella McCarron, was a close friend of Saro-Wiwa and proved instrumental in raising awareness about the Ogoni struggle in Ireland. Ogoni Solidarity Ireland began in 1994 and campaigned from Ireland on behalf of the people of Ogoni and Ken Saro-Wiwa. Despite international pressure, Saro-Wiwa and nine other members of MOSOP were hanged in November 1995 by the Nigerian military government. Five years later, the 'Shell-to-Sea' campaign began on the west coast of Ireland opposing the development of an onshore gas refinery by Shell.

3

Revolt Against One-Worldism: Radical Claims on Land and Work Post-1968

Ten years after *Silent Spring* was published, British economist Barbara Ward helped set the intellectual scene for a new phase of environmental thinking, characterising the planet as a 'fragile orb' that could only survive if people joined together globally. Ward's vision helped define the first international conference on the environment and development, the United Nations Conference on the Human Environment, held in Stockholm in 1972, having been set out earlier that year in her co-authored book, *Only One Earth: The Care and Maintenance of a Small Planet* (Ward and Dubos 1972), and the accompanying short film, *Survival of Spaceship Earth* (IIED 2014 [1972]). The image on the book's front cover is, unsurprisingly, the 'earthrise' image taken by astronaut William Anders in 1968 as part of the Apollo 11 space mission (see Interlude 2). Both the book, which was commissioned by the newly established United Nations Environmental Program, and the film evoke the shared biophysical, earth system as the basis from which to forge a truly global community. 'Spaceship' earth is our home, we learn, a web of fragile connections that needs protection. It is this planetary call that demands global unity and collaboration.

An eminent white academic, Ward provided a rallying call that galvanised social and political action, just like Carson's did. Yet the new imagery was problematic in related and in new ways. While emphasising justice in her cooperating planet – this newly forming global community requires a more equitable use and allocation of resources, between rich and poor, she argues – the planet itself is presented as unifiable, universalisable and whole. Meanwhile, although Ward references the frustrations of a younger generation fed up with the 'blatant forms of materialism' and the inability of established institutions to change, the imperative to act is still placed on northern global leaders and white scientists (of the 'First World' in the 1970s). This vision is markedly disconnected with the aesthetics and claims of mass social and political movements organising against war, colonialism, patriarchy and industrial capitalism in the same period. How and why did environmentalism base its vision in one shared world, when so much of that world was insisting that this unified world did not exist?

In this chapter we focus on the emergence and fracturing of what we call 'one-worldism' between 1968 and the mid-1970s, at the same time highlighting how movements beyond it offer resources to environmental thinking that were only partially incorporated in that moment. One-worldism refers to the way that the earth was imagined as the shared physical and scientific foundation for a new global community in accounts like Ward's, leading to shifts in the organisation of global environmental protection.

Although, in many ways, globalisation, global trade routes and global communications had been in place for centuries, new scientific and technological innovations and forms of material infrastructure intensified these connections after WWII, giving rise to new institutions.[1] Moreover, the extra-planetary view of the earth ('earthrise') made possible by space travel and fresh understanding of the world's interconnected ecologies made it possible to talk in terms of planetary agendas and policy making, as epitomised by Barbara Ward's invocation of 'spaceship earth'.

Calls for wholeness and unity often surface in times of dissent and conflict. It is no coincidence that the idea of *global* unity in environmentalism emerged at a time of intense social unrest. The United Nations Environmental Program conference in 1972 may have advertised itself as an international meeting, but the entire Soviet bloc was missing: this was the height of the Cold War and conflicts were hot across Latin America, South-East Asia and the African continent. Meanwhile, many newly independent nations attended such meetings on their own terms for the first time, beyond the banners of empire.[2] Outside the conference, disenfranchised youth, feminist, anti-colonial, and anti-imperial movements, the Black and Civil Rights Movement, and militant labour movements were emerging in force. Within industrialised countries, today known as the Global North, the refusal of work (and the alienation it gave rise to) marked the starting point for radical reimaginings of autonomy and collective life beyond capitalism. The late 1960s and early 1970s was a time of conjoined political, economic and environmental crises that threatened the foundations of liberal, post-war capitalism (Nelson 2015). While such growing transnational movements were not primarily focused on environmental issues, themes of land and livelihood grew, especially as awareness spread of the threats posed by techno-scientific scales of production, and new forms (and geographies) of neocolonialism.

In this chapter we follow how these social and political movements beyond environmentalism nourished and politicised environmental debates. In contrast to the one-worldist ideas growing in international environmental forums, the critiques and practices of these movements insist that ecological reparation is only possible when global hierarchies of imperial power and capitalist accumulation are dismantled. This, however, requires a break with one-worldist imaginaries, to understand that more than one world is at stake. An incipient Third-Worldist movement made this point in political terms during the 1960s, although later the political force would be drained as the terminology became appropriated by humanitarianism and international aid. When environmentalism becomes disarticulated from these far-reaching structural critiques and internationalist movements it becomes impoverished. In later chapters, we will build on this argument to show how these early concerns with decolonisation, labour and social reproduction also set the scene for rethinking environmental protection in more holistic and deep-rooted ways. This work also contributes to broader genealogies of environmentalism because, while we often assume that environmentalism 'went global' in the 1990s, we highlight the period 1968–1973 as a critical moment in transnationalising collective thought about the environment, and in questions over how the movement would be transnationalised. While the radical, Third-Worldist and dissenting movements connected globally to form critiques and alternatives, the one-worldist vision came to be the dominant one by the end of the 1970s. It is vital we observe how this happened, and what becomes lost when these wider movements and claims become disassociated.

The remainder of this chapter is structured as follows. We begin with the ferment that followed the 'failed' revolution of 1968, in Paris, but also in a wide range of newly connected contexts. Emerging circuits, spaces of connection and movements contain the radical energies objecting to the project of one-worldism. The next section focuses on worker and feminist movements in parts of Europe and North America that reframed the politics of work in overlapping, but distinct ways. What is vital about the new politics around work in this period is that the concerns, sites and demands of workers (including women as reproductive workers) shifted from wages and the factory floor (or mines) to the wider community, health and environment, breaking down any neat distinctions between work and life (Barca 2019). Third, we follow how anti-colonial movements across Africa, the Caribbean and beyond established their own visions of livelihood beyond ideas of European 'progress' relying on the continued extraction of resources, exploitation of labour and domination of colonised peoples. While these anti-colonial and decolonial movements developed in different contexts, they represented a world-changing, majority impulse for a new horizon beyond capitalism and imperialism – as Vijay Prashad puts it: 'The Third World was not a place. It was a project' (2007: xv). We explore the ramifications of this project through the contributions of anti-colonial militant, writer and agriculturalist Amílcar Cabral, and Sylvia Wynter, a figure known for writings about colonialism and its legacies, who sought to ground anti-colonial politics in the reclaiming of the 'plot', a space apart from the metanarrative of European humanism and the naive appeal of pre-colonial culture.

1968 and the world problematique

In May 1968, a period of civil unrest across France culminated in national strikes, demonstrations, and the occupation of universities and factories. Such protests were linked with movements taking place around the world. According to a survey Watts (2001) cites from *Le Monde*, over 70 countries had major student actions in 1968. Although it is widely seen to have 'failed' as a revolution, in that France subsequently reverted to more conservative forms of rule, the protests themselves embodied new connections between movements and geographies. Collaborations were forged between anti-colonial movements, migrant movements, labour unions and agricultural organisations. This global moment opened radical debates over the conditions of life, labour and land in ways still important today. Here we explore these debates in the context of what became known by the Club of Rome as the 'world problematique'.[3]

The late 1960s also marked a crucial period for decolonial thought in the wake of 25 years of national independence movements since the end of the Second World War (Prashad 2007). The public demonstrations in Paris and elsewhere were accompanied by currents of anti-colonial organising carried by young activists who had travelled to the metropolitan centres of Europe to study and organise against imperial powers. The Algerian War of Independence (1954–1962) had recently ended, with intellectuals in France becoming important voices in the broader questions of decolonisation and Third World development. The Vietnam War (1954–1975) also brought together global anti-war movements, consolidating fresh arguments against the intervention of countries like the US abroad, while the Chinese Cultural Revolution (1966–1976) led by Mao Zedong, Chairman of the Communist Party of China, was a further influence for ideas about leaders 'getting off their horses' and becoming involved in struggle rather than directing from a distance.

Scholars like Michael Watts (2001) and Kristin Ross (2008) observe how the concept of the 'Third World' emerged at this time as a consolidation of the critique of the ways colonialism had, and was still, shaping the contours of the new world. Ross (2008) notes that this 'Third-Worldism' emerged as a political concept, connected with the aims of self-determination, progress, and rights to land and livelihood within a restructured global economy.[4] Although it would take on the more ethical tone we associate with it today from the late 1970s, becoming associated with pity politics, and a language of victimhood, at its inception it was meant as a critique of the historical forms of imperial privilege enabling systematic forms of exploitation of the global poor.

A key event in the formation of the Third World project was the Bandung Conference of 1955 in which 29 countries from Asia and Africa (representing 1.5 billion people, or 54 per cent of the world's population), met in Bandung, West Java, Indonesia, to promote Afro-Asian economic and cultural cooperation in opposition to (neo)colonialism. The Bandung Conference, linked with the rise of the Non-Aligned Movement (NAM) from 1961, can be seen as a precedent for alternative globalisation summits organised ever since. The NAM embodies a major project of delinking with the one-worldist project. Around the world and among diasporic intellectuals engaged in emerging collectives and movements, the question of how to form an alternative international economic order was ripe and urgent.

In Paris, the student protests remained the most vociferous site of exchange and articulation, addressing (neo)colonialism, corruption, African state politics and gender inequalities (Slobodian 2008). Such protests constituted new solidarities between political groups, mobilising the resources and claims of foreign students into mass campaigns, such as the regularisation of citizens brought to France on temporary visas to construct its 'empire' (Feldblum 1998). This mobilisation was facilitated precisely by the growing internationalism of the student body. A rising presence of students from the Third World in European universities (in Germany numbers rose from 200 in 1951 to 12,000 by 1962, partly financed by state scholarships, according to Slobodian [2008]) fostered an awareness among European students of global conflicts and the foregrounding of colonial issues in political organising. New intersections between student and labour movements with migrant-led movements also fomented new strategies and solidarities in and beyond Europe (Germain 2008). Alongside meetings, journal printing and the occupation of buildings, these activities expanded to include the vandalisation of African embassies and sit-ins in solidarity with independence struggles.[5] Watts (2001) links the emergence of this collaborative proto-internationalism with a distinctive 'Southern' moment focused on moving beyond oppressive structures of policing and state control.

Anti-colonial movements were also becoming an important influence on emerging environmental movements. Between 1968 and 1973, the Brazilian geographer Josué de Castro taught in the University of Vincennes, the radical college built on the outskirts of Paris in the immediate aftermath of May 1968. His classes encouraged younger scholars to experiment with new political approaches to human ecology, appealing for greater academic and political effort to connect the growing ecological crisis with Third World underdevelopment (Davies 2022). De Castro also led a research group focusing on social and ecological transformations in the Amazon (Davies

2022). Written up into a report in 1972, the intention was to gain greater publicity for the scale of damage, as well as proposing solutions that could feed into an international response (for example, through the forum of the 1972 Stockholm Conference).

The linking of environmental degradation, underdevelopment and (neo) colonial structures is also traced by Lazure-Beecher (2018) in the context of Québec, Canada, through the prism of a radical journal (*Partis Pris*) established in 1963. In retelling the historical narrative of this moment, Lazure-Beecher notes the importance of voices from the decolonising world to ideas and practices in the emerging environmental movement. Early environmentalism tends to be credited to the intellectual work of American intellectuals, while 'Third World' scholars are primarily associated with the project of anti-imperialism and decolonisation. However, this work reveals the importance of African and Caribbean intellectuals: Aimé Césaire, Léopold Sédar Senghor, Albert Memmi, Frantz Fanon, and others on environmental thought, albeit in currents that became sidelined in subsequent decades.[6] Members of *Partis Pris* joined the environmental movement in the early 1970s for such political reasons, seeing in it an opportunity for 'revolution', especially in the wake of other failures. Centres of thought like *Partis Pris* can be seen as a kind of (anti-colonial) alternative to the dominant form of environmentalism consolidating in North America at the time.

Building on the connections between ecology, critiques of capitalist political economy, and social movements, in the next section we turn to the work of André Gorz, a Paris-based thinker who connected better than most the struggles around work and ecology taking place in the wake of 1968. This was a period when the horizons of social transformation were radically expanded, in large part due to the extension of sites of struggle beyond the confines of the factory. We look at how workers organising in some sectors challenged the capitalist value system that determined what was being produced and how this linked to broader social concerns with the ecological effects of industrial capitalism. We also look at the vital work of feminist activists and movements that challenged the international capitalist division of labour, institutionalising hierarchies of work/value and masking the exploitation of unwaged work – an intervention that created space not just for women, but for all subaltern subjects undervalued within the world-capitalist system (see Barca 2020).

'Proletarians of the world, unite!': work, social reproduction and ecological crisis

In 1974, Sicco Mansholt, former president of the European Common Market, stated in an interview with *The Ecologist* magazine: 'Ever since

1968, I have been convinced we have to stop growth not only for ecological reasons but because of the enormous gap between developed and developing countries' (Mansholt and Walker-Leigh 1974: 259).[7] That a leading European politician would advocate for a no-growth economy in the early 1970s says a lot about how extraordinary this period was, and is a testament to the social and environmental movements that had been organising over the previous decade.

Yet Mansholt is perhaps most famous for the so-called 'Mansholt Plan' (1968), which ushered in massive reforms of the European agricultural sector. These included reducing price supports for farmers and the rationalisation and mechanisation of farming operations. These new proposals fuelled the anger and frustration of tens of thousands of farmers across Europe who had also taken an active part in the social mobilisations of 1968. This culminated in 1971 with between 80,000 and 100,000 farmers filling Brussels city centre to protest against structural reform, and falling farm prices and incomes. Effigies of Mansholt were hung up and placards read: 'Mansholt – the death of peasants'. Farmers threw cobblestones, firecrackers and rotten eggs at the police forces, who reacted with tear gas and water cannons (Germond 2015). Mansholt's response to the farmer protests was that such animosities arose from narrow economic self-interest and a lack of understanding of the wider issues and context (Mansholt and Walker-Leigh 1974). He also comments in *The Ecologist* interview that there is no sign of a 'popular ecological movement' in Europe because only a fraction of the population is 'ecologically aware' and that any ecological policies are bound to be unpopular.

Mansholt's quick dismissal of the farmer protests and support for market reform and technological efficiencies aligns with dominant strands of environmental thinking in the 1970s – for example, the apocalyptic warnings of Garret Hardin, Paul Erhlich and the 1972 OECD *Limits to Growth* report. While of different political persuasions, these positions share a technocratic, austerity-focused understanding of the ecological crisis, emphasising greater technological efficiency, individual sacrifice and better regulation of surplus populations, but no fundamental transformation of capitalist over/under production and consumption.

Pushing against this grain were, we argue, a host of related but far from uniform struggles that did not identify as 'environmental' but which engaged with fundamental questions over the nature of work under capitalism, exchange value versus use value, social reproduction, technology and collective freedom. As such, they politicised the home, family life, sexuality, consumption *and* the environment (Federici 2019). André Gorz, an Austrian-French writer, was one of those making connections between these different domains of social and economic life. Gorz moved to Paris after the Second World War and was part of the New Left intellectual culture that flourished there, being close to Jean-Paul Sartre, Herbert Marcuse and Ivan Illich, thinkers seeking to rethink freedom amid transforming institutions and political conditions. With the anti-colonial thinkers discussed later in the chapter, Gorz's link to environmental politics stems from his concern with the 'total forms of frustration' (1980 [1975]: 4) felt by so many sections of the population and his effort to liberate multiple aspects of lifeworlds, including ecological aspects, from domination. This stands in stark contrast to Sicco Mansholt's pathologisation of these forms of dissent.

More than most of his contemporaries, Gorz foregrounds environmental problems as part and parcel of a wider crisis in post-war capitalist development (Leonardi and Benegiamo 2021). In the early 1970s, and way ahead of his time, Gorz was quick to recognise that what was being called the 'limits to growth' could be exploited to enable capitalism to continue expanding its own logics of private accumulation. Ecological crisis, in other words, spelled new opportunities for capitalism, rather than marking its end. Thus, he predicted, business elites would reorganise to ensure their activities would continue, specifically by forcing out competitors with less resources to 'innovate', or by outsourcing polluting activities to the Third World. With prescient foresight, Gorz anticipates the new 'green' products and technological innovations that would follow this incorporation of the new limits, sketching the outlines of the business-friendly, technocratic frameworks of sustainable development and ecological modernisation we see multiplying today.

Post-war economic growth had been premised on selling a dream of prosperity and success through the expansion of waged work and the consumption of mass-produced, disposable commodities (Packard 1960). By the 1960s, this dream was coming apart at the seams, as technological promises failed and sites of protest multiplied. Against the disposability and planned obsolescence of post-war industrial capitalism, and its socially and ecologically destructive consequences, Gorz argued that technological innovation could better be directed towards social value and shared needs, rather than economic growth and profit-maximisation. This would reduce the time required to carry out dull and harmful work, expanding individual autonomy to pursue diverse interests.[8] This is not the abstract liberal freedom of individual rights and the market, but individual freedom supported through public and community infrastructures, such as mass transit, open source knowledge and shared, communal resources. It included housing consisting of private and common spaces, workshops and libraries, playrooms and laundry areas; universal access to shared tools, machines, materials for making and repairing everyday necessities; and public transport that would remove the need for private cars.

Gorz's own thinking was influenced by and reflected broader currents of thought being articulated by feminist and worker-led movements. Traditional labour demands were changing. In 1966, Mario Tronti, a leading figure in Italian autonomous Marxism, wrote a pamphlet titled 'The struggle against labour'. In it he asks: 'What then of labour as "the prime necessity of human existence" (Marx)? Perhaps it would be better to transfer it from the future prospect of communism to the present history of capitalism – to let the workers drop it and consign it to the bosses' (Tronti 1966: 25). Living better, not earning more, was where some worker demands were moving, and living better depended 'less and less on individual consumer goods the worker can buy on the market, and more and more on social investments to fight dirt, noise, inadequate housing, crowding on public transportation, and *the oppressive and repressive nature of working life*' (Gorz 1980 [1975]: 133; emphasis added). The rejection of work, of the identity of the worker, was profane for many Marxists, traditional unions and left parties. Yet posing questions about the value of work and the nature of industrial production made space for associating political demands.

Importantly, questions concerning the politics of work were being linked to wider ecological concerns through such inquiry and activism. In Italy, petrochemical workers in Porto Maghera, organising in the 1970s, challenged both the quantitative aspects of work-time (how many hours, for what wages) and the qualitative dimensions of production (what are we producing, under what conditions).[9] As Feltrin and Sacchetto (2021) have shown through their archival work on the Porto Maghera workers, the 'noxiousness' of chemical production was linked by the workers not only to their own health and the health of the surrounding community, but to the entire planet – the mass production of PVC plastics in the refineries of Porto Maghera was understood as being harmful at scale. The only rationale for this type of production was the imperative of exchange value and profit-maximisation within a capitalist system. Against the 'noxiousness' of this fossil-fuel based industrial complex, how could the nature of work be transformed, they asked collectively. What is so significant about this line of questioning is that it articulates work, value and ecology together, and that this articulation emanates from workers themselves. It both challenges the fundamental organisation of capitalist work and opens the possibility that workers, liberated from capitalist work, could collectively design, make and share products based on use value, rather than exchange value. Although autonomous forms of organising in Northern Italy have had considerable influence on (Marxist) political thought, the connection with environmental thinking has not often been drawn out (see Barca 2019).

Social critiques of industrial technology were widespread at this time – reflecting a popular fear and uncertainty around new technologies (see Chapter 2), but also a demand that technologies be developed for the social good, to foster human flourishing. This was particularly evident around the mass, transnational anti-nuclear movement (Gottlieb 2005). Less well-known were the demands put forward in the Lucas Plan by shop stewards working at Lucas Aerospace Corporation in the UK.[10] Faced with closure and the loss of thousands of jobs, the workers devised an alternative plan for the company that included the manufacture of 'socially useful products' (Lucas Plan 1979).[11] In their clearly worded 'Corporate Plan', the shop stewards' committee placed their plan within a wider context of what they called 'industrial democracy', echoing the *qualitative* aspects of work that Porto Maghera workers had emphasised, noting that 'trade unionists are attempting to transcend the narrow economism which has characterised trade union activity in the past and *expand out demands to the extent of questioning the products on which we work and the way in which we work upon them*' (Lucas Plan 1979: 9; emphasis added).

What was so radical about these examples of labour politics is that they extended more familiar worker concerns (with wages, with fair conditions) to broader social and environmental concerns. Barca and Leonardi call this 'working-class ecology': 'the web of systemic relations between working-class people and their living and working habitats' (2018: 489). And it was not just waged workers or unions questioning the foundations of the capitalist political economy – what was being produced, why and by whom. Within the feminist movement, radical claims around work and value were also being made. Feminist theorisations of social reproduction argued that value within capitalism is not only produced by waged, 'productive' labour, but by a much larger field of unwaged, *reproductive* labour: the work of carrying, birthing and caring for children, of feeding, supporting and taking care of the predominantly male workforce.

Social reproduction, as the feminist thinker and activist Silvia Federici writes, was not something that feminists discovered in the 1960s; liberal and Marxist economists had long discussed the ways society reproduces itself (2019). What developed in the 1960s and 1970s, through the work of feminist theorists and activists, was an awareness that labour exploitation under capitalism does not just happen via the wage relation, but that it extends to the work of millions of unwaged house-workers and many other unpaid and un-free labourers. This argument broke down the distinctions between domestic and work-places, private and public, illuminating the (re)productive work carried out by women in the household but also opening space for the intersection of allied struggles. This was the basis of the International Wages for Housework Campaign which was not focused on women's rights or access to the workplace for women, but demands for payment for domestic labour – and as such, the destruction of the hierarchical (patriarchal) capitalist division of labour (Federici 2019).

Two of the leading figures of the International Wages for Housework Campaign were Selma James, an American writer who was steeped in Black liberation and international, anti-colonial movements (living part of her life in Trinidad) and Mariarosa Dalla Costa, whose political experience was shaped by the Italian expressions of autonomist Marxism, workerism[12] and the student revolts of the 1960s. The two met in London in 1971, going on to publish the seminal feminist text *The power of women and the subversion of the community* a year later (Bracke 2013). The same year saw the launch of the International Wages for Housework Campaign with branches in Italy, the UK, the US, and later other parts of the world. The work of these feminist thinkers does not stop at linking capitalism with the exploitation of women's work, but extends critique to the institutions that inscribe this division of labour at an ideological level, including among left organisations, education, the family, sexuality, the body (Dalla Costa and James 1997) (see also Box 5). Their work has also continued to inspire future generations of feminist thought and activism, particularly as it has become more explicitly linked to ecological questions via ecofeminism (Bennholdt-Thomsen and Mies 2000), feminist science (Haraway 1991) and conceptualisations of more-than-human care work (Puig de la Bellacasa 2017).

Illustrating the cross-cutting eddies and currents of intellectual thought and politics at this time, the International Wages for Housework Campaign linked trans-Atlantic feminist movements and the movement for Black liberation – the term 'racism' was translated into the Italian situation by some of the early feminist campaigners to express women's discrimination and male privilege (Bracke 2013). Nor can the feminist campaign be separated from the revolts of the unwaged proletarian populations of the European colonies. In *The wretched of the earth*, Frantz Fanon writes: 'This European opulence is literally scandalous, as it has been founded on slavery, it has been nourished with the blood of slaves and it comes directly from the soil and subsoil of that underdeveloped world. … The wealth of the imperial countries is our wealth too' (2001 [1963]: 79–81). From the factory to the household, the metropole to the field, new stages of social antagonism were surfacing as the 'subliminal "other"' (Salleh 2010) exploded the category of working class. The working class was no longer just those organised by the unions and left parties, or those who received a wage for work-time, but the class of the exploited who, now, refused to be exploited through their labour. 'Proletarians of the world, unite!' was Mario Tronti's resonant call in 1966. What united the proletarians was a refusal of subjugated work in general, the work carried out beyond the factory gates. If the category of work expands, so too does the terrain and circulation of struggle – here is the real source of the 'world problematique' and the radical ecology of working-class politics at this time.

Box 5: Zerowork and the trans-Atlantic connection

It is not always easy to pinpoint moments or spaces of encounter and connection that shaped new ideas and social movements. In the 1970s, one of these spaces was the collective Zerowork. Formed in 1974 in North America by a group with diverse intellectual and political backgrounds, Zerowork came together to form an analysis of the overlapping local and international crises they saw emerging around them, with an emphasis on labour, and implications for workers' struggles. Influenced by new currents of Marxist theory, particularly Italian autonomism, as well as feminist and anti-colonial writing, individuals involved drew on activist experience on both sides of the Atlantic in their analysis, as well as the different movements they were involved in, including the Civil Rights Movement and the International Wages for Housework Campaign. The individuals who made up Zerowork – including Harry Cleaver Jr., Peter Linebaugh, Silvia Federici, Mariarosa Dalla Costa and Raniero Panzieri – were dissatisfied with dominant explanations of the 'world problematique' (see note 3), including from traditional leftist positions, and sought to open a new analysis, developed through the creation and publication of the *Zerowork* journal. The writings of Zerowork capture a desire to connect the multiple sites and forms of struggle around work that had been erupting and circulating since the mid-1960s.

Harry Cleaver (nd), one of the key figures (and archivists) in Zerowork, outlines how this many-headed social movement consisted of different struggles connected through a refusal of capitalist hierarchies of work (hence 'Zerowork'). Waged workers were fighting for more collective bargaining rights, for more money, better working conditions and fewer working hours. Unwaged workers included women objecting to the capitalist division of labour, as well as students fighting for free speech and an education system that reflected their desires. The collective also included Black and Brown militants fighting for civil rights and racial equality; prisoners (disproportionately Black or Brown) fighting for greater legal rights and more freedom to study and communicate, and peasants fighting for land, for autonomy and for liberation from foreign domination, whether colonial or neocolonial.

Zerowork understood all these struggles as connected, part of an extended working class that included the waged and the unwaged, a transnational movement that could unite across their differences to bring about radical social change. It was thus as much a critique of traditional left ideas and organisational strategies as it was a critique of capitalism: in seeking to overthrow capitalism, the old left was simply unable to accommodate the new circulation and energy of these proletarian, transnational revolts. As Mariarosa Dalla Costa, one of the founders of the International Wages for Housework Campaign, writes: 'To the extent to which capital has recruited the man and turned him into a wage labourer, it has created a fracture between him and all the other proletarians without a wage who, not participating directly in social production, were thus presumed incapable of being the subjects of social revolt' (Dalla Costa and James [1972] 2019: 77). This quote captures well the explosive, if complex and uneven, arrival of new political subjects on to the stage of history in the 1960s and 1970s – women, colonial subjects *and* nonhuman natures (Bennholdt-Thomsen and Mies 2000).

Radical claims on land: anti-colonial struggles and creative connections

At the end of *The Wretched of the Earth*, Fanon (2001 [1963]) addresses his anti-colonial comrades, warning them not to draw inspiration from Europe in their quest to decolonise Africa – a direction destined to culminate in 'obscene caricature'. What he means by this is that the project of decolonisation needed to take place on very different terms: it needed its own forms of imagining, social organising and aesthetics.[13] Fanon portrays the colonial system as one concerned with certain forms of wealth and resources in its ecological endeavours, to the exclusion of the soil and the earth. 'Perhaps it is necessary to begin everything all over again', he concludes, and 'to re-examine the soil and mineral resources, the rivers, and – why not? – the sun's productivity' (2001 [1963]: 79).

Anti-colonial and Third-Worldist demands are frequently premised on claims about land, ecologies and livelihoods, and yet are rarely considered as environmental struggles.[14] For Wenzel (2017), Fanon exposes how marginalised peoples' struggles over nature and for justice have long been waged in an idiom, quoting Martinez-Alier, 'that is not explicitly environmental but nonetheless engages the concerns of environmentalism' (Martinez-Alier 2003: 169). In like manner, postcolonial theorist Pablo Mukherjee (2010: 25) argues that, if 'greening postcolonialism' seems a strange project, this is only because scholars shaping environmental thought in the 1980s overlooked the high importance on 'land, water, forest, crops, rivers, the sea' in earlier anti-colonial thought.[15] Thus, the issue is twofold. First, anti- and decolonial thought and praxis are discounted from histories of environmentalism, while the environmental dimensions of anti-colonial movements and decolonial theory are likewise underplayed.

Anti-colonial groundwork

One of the most interesting individuals involved at the intersection of agrarian and anti-colonial movements was Amílcar Lopes de Costa Cabral, a Bissau-Guinean and Cape Verdean agronomist and one of the leading revolutionary figures in African anti-colonial struggle (see Chabal 1983; Dhada 1993; Wood 2016; César 2018). Like other anti-colonial and decolonial thinkers of this period, Cabral undertook his studies in the colonial core, studying agronomy in Lisbon. Rather than dispelling his desire to rid his home country of Portuguese rule, the experience deepened his political education, particularly through his involvement in antifascist student activism (Manji and Fletcher 2013). His ability to strategically select, combine and apply different European and African intellectual traditions, including Marxian political theory, agricultural science and world history, is a hallmark of his distinctive contribution to anti-colonial ecological thought (Ferretti 2021).

On graduating in 1952, Cabral turned down an assistant professorship at the Instituto Superior de Agronomia in Lisbon, instead moving back to his country under a contract for the Portuguese Ministry of Overseas. He and his wife, Maria Helena, lived in the house assigned to the director of the Experimental Farm of Pessubé. This farm had been used to produce vegetables for the colonial political and administrative elites of the city, as well as serving as a recreational space for picnics and recreational walks (Schwarz 2013). Cabral transformed it into a centre for agricultural research: a means to improve and modernise the production of Bissau-Guinean farmers. This included identifying new growing techniques, new crop varieties and local strains of commercial crops (for example, jute). As a deputy in the Agricultural and Forestry Services, Cabral was also tasked with carrying out the first agricultural census in Guinea-Bissau (McCulloch 2019). The fieldwork and survey took three years to complete, providing Cabral extensive and prolonged contact with the agricultural economy and rural population of Guinea. Through this work he was also able to trial some of the techniques and crops developed in Pessubé.

The experience of travelling around Guinea-Bissau, learning from farmers and the constraints under which they operated, was formative for Cabral: with over 30 per cent of the cultivated land devoted to the production of groundnut for export, he saw first-hand the negative impact of colonial monocultures on land and people. He also identified the different social strata, ethnic groups, land-types and farming techniques across the country (Schwarz 2013). Cabral saw in all this the central role that land and agriculture must play in national liberation and the anti-colonial struggle.

Cabral thus became committed to decolonising Cape Verde and Guinea-Bisseau through his engagement with land-based issues and agronomy.[16] Between 1963 and 1973, the year he was assassinated, he led the Partido Africano da Independencia da Guine e Cabo Verde guerrilla movement in Guinea. Cabral trained his soldiers to work with local farmers to develop better farming techniques so that they could better feed their own communities, as well as Partido Africano da Independencia da Guine e Cabo Verde soldiers (Gibson 2013). While being an intellectual and organisational driver of the liberation movements in Guinea-Bissau, Angola and beyond, he developed novel approaches to soil repair, setting up experimental farms where new ideas could be tested. Resonating with the agroecological approaches developed by Mexican agronomists in the 1960s, as well as the work of fellow Lusophone Josué de Castro, Cabral remains a pioneer in terms of his understanding of the complex interplay between farmers, technology, soil, plants and climate. He can also be seen as an early proponent of food sovereignty, the concept expressing the rights of post-colonial nations and of peasant communities to choose how and what to grow (see Chapter 3), emphasising the centrality of the farmer in maintaining the productivity and health of the land – in contrast to ideas of farmers as 'ignorant' and 'anti-environmental' still expressed in developmental discourses today (Schwarz 2013).

While modern environmentalism has tended to imagine the 'environment' separate from human activity, Cabral joins here with anti-imperialist movements and critical scholars like Fanon and Aimé Césaire who consistently emphasise this veil of separation as a key mode for justifying the displacement of people from conservation areas (Wenzel 2019). Thus, for Cabral (2016: 261), soil is 'the inscribed body and erosion is the scar left by historical violence', while for Fanon, land should be understood as 'a primary site of postcolonial recuperation, sustainability and dignity' (1961, cited in DeLoughrey and Handley 2011: 3). For anti-colonial thinkers, soil was not only tied to questions of chemical toxicity, becoming tangible as a pollutant on animal and human bodies – although, as we saw in the previous chapter, this was relevant. Rather, soil was connected in a primary way to dispossession and alienation, a broader kind of 'poison', foundational in creating the conditions of industrial development in the imperial core, and de-development in the peripheries. Influenced by the lessons of his father, Cabral considered that the droughts he encountered in Cape Verde were not merely 'natural' but largely a result of colonial policy (Wood 2016). Cabral's emphasis consequently falls on improvising practices that can transform and restore soil health *and* alter colonial power dynamics.

Reasserting the centrality of peasant farmers within food production was never, however, just about reorienting agriculture's priorities. Because the cultural resources of the Indigenous population remained strongest in rural areas, land and agronomy centrally concerned a reassertion of the agency of Indigenous people. Cabral's emphasis on culture and its role in national liberation is, like Fanon's, not to be interpreted as an unconditional affirmation of 'African' culture as a pre-existing set of traits or traditions – indeed, it was precisely this kind of binary thinking that he suggests might prevent Africans from embracing the rapid changes required to become part of 'humanity' (Cabral 2016). Instead, he calls for the forging of a new historical trajectory that involves culture as an integral part of social and economic progress, without emulating the West, or returning to an imagined past. Identifying with the people was not just a rhetorical form of populist opportunism, but an ongoing *praxis* undertaken through 'daily contact with the mass of the people and the communion of the sacrifices which the struggle demands' (Cabral 2009 [1970]). Drawing on agricultural metaphors as he did, Cabral understood culture as both the root and fruit of the struggle for national liberation. With decolonisation and self-determination, a new mode of production most appropriate to African development would 'necessarily open up new prospects for the cultural process of the society in question, by returning to it all its capacity to create progress' (Cabral 2009 [1970]). Bringing experimental technical agronomy and participation in nationalist revolutionary struggles together was, for Cabral, the basis of anti-colonial 'groundwork', encompassing care not only for land, flora and fauna but for people and their wider social relations.

Plot and plantation

Across the Caribbean, the project to transform relationships with landscape as part of a project of decolonising relations with land and soil was taken up by a range of intellectuals, artists and activist-thinkers, including Aimé Césaire, who drew on botanical history to evoke African and Arawakan roots, and Sylvia Wynter, a transdisciplinary scholar situated between the decolonising cultures of the early postcolonial and post-Civil Rights Movement between the 1950s and late 1970s (Rodriguez 2015). Having moved between the decolonial intellectual communities of the Caribbean archipelago, Britain and the US, Wynter embodies a form of social inquiry 'created by structurally subordinated peoples … between colonies and metropoles' (Rodriguez 2015: 7) reframing emerging debates like environmentalism by creating new connections. Wynter's thought is particularly pertinent for this chapter because it speaks to diasporic centres and margins of colonialism, revealing contradictions in the ways that modern environmentalism solidified post-1968 largely to the exclusion of anti-colonial and anti-racist movements.

Like Cabral's contrast of peasant agriculture and colonial systems of food production, Wynter's early work distinguishes between the folk 'plot' of islander agriculture, and the colonial 'plantation' (Wynter 1971). The plantation economies and ecologies imposed one crop in large monocultures – which DeLoughrey (2011: 59) writes of as 'Euclidean grids' – and were farmed using indentured or slave labour to produce food supplies for growing European consumption. In contrast were the forms of land cultivation through which the islands' inhabitants customarily grew the food crops, including yams, cassava and sweet potatoes – crops that were not of interest to colonial incomers. The accumulation of space for plantations as well as agricultural techniques involved transforming the botany and social cultures of Caribbean islands in violent and lasting ways. Most significant in this history was the establishment of the sugar plantation, which required large-scale clearance of dense and moist tropical forests, and displaced pre-plantation Arawak subsistence practices of agriculture and fishing. The exploitative expansion of large-scale sugar cropping led to soil epidemics as well as landslides, erosion and climatic changes linked with rapid deforestation (Paravisini-Gebert 2005). Meanwhile, African slaves were transported to produce sugar and other luxury crops such as coffee, rum, spices and bananas for the international market.

While the plantation/plot narrative device primarily documents this violence, Wynter also explores how the latter might claim and recast a political and socio-environmental future *beyond* colonial coordinates. With others like Glissant, Walcott and Kincaid, Wynter defends the role of literature in forging an environmental imagination in the Caribbean with anti- and decolonial possibilities. While her literary work and essays do not often address environmentalism *per se*, Wynter's project,[17] with Cabral's, exposes the role of coloniality in ecological degradation, and the necessity of land and labour justice for any kind of repair.

The violence involved in the making of plantations was explicitly racist, and led not only to dispossession and ecological degradation, but to the establishment of degrading/appropriative ways of seeing people and their practices. As anti-colonial and postcolonial thinkers have long emphasised, Caribbean geographies were a central part of colonial imaginations and trade routes that cast a shadow on the islands' ecologies, economies and people, long after nations achieved independence. The islands were imagined as hot, tropical paradises of lush vegetation, alternately portrayed as tropical escapes, 'womb-like' places of primordial fertility and sources of amoral danger (DeLoughrey et al 2005; McKittrick 2013; King 2016). Within this colonial imaginary, Black and Indigenous bodies were frequently portrayed as plants, vegetation and spatial potential – for example, in travelogues of the 17th century where the islands were scoped for their potential to support the political economies of Europe (King 2016; Fehskens 2019). To speak of the 'hybridity' or 'creolisation' of culture that occurred in this new international ferment without attending to this violence is a gross romanticisation. However, as Wynter, Glissant and others make clear, the tremendously energetic imagination that characterised literary, poetic and theoretical imaginations in the wake of plantation violence did draw on the new transnational connections between oppressed and dispossessed peoples. This transnational connection is sought and constructed on very different terms.

As the basis for an anti- and decolonial future, the 'plot' Wynter (1971) evokes is the ground for imagining solidarity, connection and creation in terms that do resonate transnationally, but not via the vocabulary of the coloniser. The plot refers to the agricultural practices but also folk imagination, stories and ways of doing that were used by Carib and Arawak peoples before colonisation – their 'indigenous, autochthonous system' (Wynter 1971: 96). However, it also refers to the resonant practices of the newly arrived African peoples, uprooted through slavery, albeit manifesting different folk customs and imagination. Meanwhile, the figure of plantation comes to anchor a series of debates about racism and the knotted-creolised organisation of diasporic life in the new world – 'knots emerging in plant and culinary life, representational politics, transnational socioeconomic linkages, black Atlantic time-space, and more', suggests McKittrick (2013: 4). This is one reason the plantation remains an important contextual setting for many great works of Caribbean fiction, a backdrop of 'official history', against which 'secretive histories' of the plot are elaborated (McKittrick 2013: 36).

This interest in devising literary registers for anti-colonial concerns that speak to different contexts led Wynter to find affinity with Indigenous poets and collectives across Latin America and elsewhere.[18] For example, she admired the magical realism of Guatemalan writer and poet Miguel Ángel Asturias, who she claims elucidates plot/plantation struggles between 'the indigenous peasant who accepts that corn should be sown only as food, and the *Creole*[19] who sows it as a business, burning down forests of previous trees, impoverishing the earth in order to enrich himself' (Wynter 1971: 96). As Niblett (2012: 6) emphasises, within Latin America and the Caribbean in particular, magical realism can be seen to 'encode the smashing of indigenous, "closed-cycle" systems of subsistence agriculture' as large tracts of land were drawn into capitalist world ecologies. Both Wynter and Asturias use this tension to identify, theorise and expand the possibility for new anti-colonial literatures. In this sense, they pre-date ecocritical literary readings by evidencing the fundamental antagonism between two very different modes of worlding ecological worlds (Paravisini-Gebert 2005).

Places and people associated with plantation spaces like those developed in the Caribbean were consequently rendered *outside* of 'humanity', thus rendering them open to appropriation, or to be sacrifice zones of environmental decay, pollution and military experimentation (Nixon 2011; Paravisini-Gebert 2005). Countering this (anti)humanist and racist system requires, for Wynter, the invention of a new humanism – a new way of inhabiting the contours of the human and making knowledge about the world. As in Fanon's thought, this anti-colonial theory claims that wider forms of degradation – ecological degradation, or deforestation, for example – cannot be thought without reworking fundamental ideas of the human. This is a deepening of the concern we have raised around modern environmentalism: that it enables labour and land rights issues to be sidelined, and colonial histories to be obscured. Here, this move is situated within longer histories of making knowledge about humans and nonhumans.

Such a disruption, aligning with Third-Worldism of the contemporaneous moment and its challenge to one-worldism, wants environmental repair and reparation on very different terms. As Wynter (1994 [1992]) would later and more explicitly write in an open letter, at the heart of the planetary environmental problem is a problem of perception and classification.

Although narrations of environmentalism's pasts have emphasised human agency– in both wrecking and restoring nonhuman ecologies – an anti-colonial history of the same period reveals the layering of the 'human' itself as a historical construction, rendering it possible, and even necessary, to expose some bodies, soils, ecologies, and so on, to devastation, in the service of protecting and nourishing others. For Wynter, it is not a matter of *scaling up* action to protect future environments (from humans), but of *recomposing the human* in order to bring about reparative action (McKittrick 2015). This project leads to exploration of how the earth might be humanised by practices neither 'traditional' nor 'modern' but both, by forms adapted and imported in the shadow of plantations and yet refusing the logic of colonisation that underpinned them (Owens 2017). As Wynter puts it, the 'single issue with which global warming and climate instability now confronts us' is the question of who is the 'referent we' (Wynter and McKittrick 2015: 24). In other words, who do news reports about environmental collapse address themselves to? Who is the human here? These remain questions of aesthetics. To reclaim the political impulse of Third-Worldism in its inception, environmentalism can't just imply there is one human community and one environment to be protected. It has to be clear that there are *multiple worlds* at stake.

Conclusion

In this chapter we have shown that there was traffic in ideas, utopian impulses, demands and revolutionary politics that travelled the rails of empire. The radical politics of the 1968–1974 period were not primarily about the fair distribution of resources or the occupation of positions of power, but a wholesale rejection of the cultural norms and logics of progress and development that had settled in the post-war period. Our central argument has been that the political emergence of Third-Worldism and decolonial politics in this moment takes issue with the basic assumptions of international organising for progress in this historical moment – what we call 'one-worldism'. Against the idea of a universal community inhabiting a 'common' planet, Third-Worldism emphasises: the historic exploitation of Third World nations and their resources for the industrialised version of progress that has emerged in the West; the need for reparations for the historical theft of land and resources; the construction of an alternative international economic order that allows for national self-determination and the levelling of historic inequalities. Third-Worldism was a rallying call to a radically alternative, technologically sophisticated and truly humanising project, capable of rescuing not just the Third World, but the West itself, from the violent anti-humanism of colonialism.

Meanwhile, the radical writings and organising of workers, feminists, Black activists and students across the Atlantic politicised waged work and social reproduction in multiple ways throughout the 1960s and 1970s, delinking work from industrial capitalism and demanding autonomy at every level – the workplace, the home, the field, the city, the school. Such demands for freedom from oppressive social structures have been dismissed retrospectively, and currently, as naive and liberal. Such 'liberatory impulses' have been dismissed for failing to organise around a revolutionary subject, and for giving rise to the individualist ethic of neoliberalism that took shape in the 1980s. But it is the transversal aspect of the new social and environmental movements borne out of the 1960s and early 1970s that we wish to emphasise – exploding familiar categories and distinctions between labour and environmentalism, home and work, waged and unwaged, production and consumption, core and periphery. This melting-pot of social critique and political experimentation offered different paths and futures for environmental politics, even if they were not taken. It is through the analysis, art and organising of activists and intellectuals that new connections were forged, new sites and subjects of struggle opened up, and new dynamic collectives formed. Far from being 'local' or 'community' focused, these ways of thinking were part of transnational currents that were the fissures cracking apart the incipient 'one-worldism' of global environmental management.

We believe that environmentalism is impoverished when it becomes disconnected from these wider claims and fundamental challenges. While the course of mainstream environmentalism was set, as we will see in the next chapter, around a notion of the environment that largely excluded concerns with work, land, patriarchy and colonialism, we suggest that the theoretical and political openings in this chapter remain opportunities available to us in the contemporary moment. Although the concept of Third-Worldism has been significantly reappropriated, the decolonial project has seen a resurgence through Indigenous thinkers and the articulation of new feminist and justice collectives who also want environmentalism on very different grounds. We return to this point in Chapters 4 and 5. Figures we have mentioned in this chapter foresaw the ways that environmental thinking without this political underpinning would enable a kind of double-think in future decades, where environmental protection for one 'global' population would come at the cost of the majority without any apparent contradiction. For example, Gorz[20] recognised that the capitalist response to ecological limits would require the doubling down on resources and sinks located in postcolonial territories. 'What a marvellous scheme!', he mocks:

> For us, clean air and water, production of non-material goods, leisure, affluence; to Third World countries, if they are well behaved,

material production, dirt, pollution, danger, sweat, and exhaustion, along with congested and polluted cities. When the Meadows report [Limits to growth] looks forward to tripling worldwide industrial production, while recommending zero growth in industrialised [*sic*] countries, doesn't it imply this neo-imperialist vision of the future? (Gorz 1980 [1975]: 85)

Second Interlude: Planetary Icons

The images that follow explore the making of the globe in and beyond Western imaginations. Like the cultural geographer Denis Cosgrove (2001), we seek to draw attention to the importance of cosmography (the depiction of the heavens) in Western planetary aesthetics. There is a tendency to think of Indigenous and non-Western imaginations of the earth as mythical, mystical and infused with cultural 'belief', while Western scientific drawings of the globe are considered objective and universal. However, the history of globes and maps in all traditions are replete with imaginings of supernatural beings, otherworldly forces and nonhuman powers (see for example Figure 10). Meanwhile, non-Western traditions have also long been interested in representing planetary holisms. Here we explore icons/iconographies that emerge through the development of particular technologies of representation and historical experiences, yet configure distinct senses of planetarity. As we explain in Chapters 3 and 4, images like *Madre Tierra*, for example, seek to cultivate an earth politics on very different terms.

Figure 10: 'The Psalter Map', c. 13th–15th century

Source: British Library, M.S, 28681, f.9, Open Commons licence

Figure 11: Persian cosmography map, '*Kharidat al-'Aja'ib wa Faridat al-Ghara'ib*' [The Pearl of Wonders and the Uniqueness of Things Strange], c. 1525

Source: British Library, MS. Or. 1525, fols. 8v-9r, London. Open commons licence

The drawing and text of Figure 11 is by Umar bin Muzaffar Ibn al-Wardi, an Arab historian. This work sums the geographical knowledge of the Arabic world of the time, referring to climate, terrain, fauna and flora, population, ways of living, existing states and their governments in individual regions of the world (Harley et al 1987). The original work is said to have been completed around the year 1419 AD, as stated on the earliest known copy which is dated 1479 AD (Harley et al 1987).[1] At the centre of the map are Mecca and Medina, the two holiest cities of Islam. The north part of the map shows 'China and India', and the south part shows 'Christian sects and the states of Byzantium'. The outer circles represent the seas. Islamic sacred geography differs from the Ptolemaic tradition in that it does not employ cartographic grids, or longitude and latitude scales; as a rule, these used Mecca and the Ka'ba as the centre of the world (Virga 2008).

Figure 12: 'Earthrise', 1968

Source: Photograph taken by Bill Anders, Apollo 8 crew member, public domain

Figure 13: 'Blue Marble', 1972

Source: Photograph taken by either Harrison Schmitt or Ron Evans, Apollo 17 crew members, public domain

A key turning point in the history of representing the earth as a globe in the West, argues Cosgrove (2001), came with the lunar view of the earth provided by Apollo 8's 'Earthrise' image (Figure 12). This new perspective produced at once 'marvel at a vast yet tiny earth, reflection on the insignificance of self, and yearning for human unity' (Cosgrove 2001: 259). However, Cosgrove suggests that this image triumphantly conveys the exceptionality of those who were able to make the world appear for the first time, providing 'visual confirmation of American democracy's redemptive world-historical mission' (2001: 260). In contrast, Apollo 17's 'Blue Marble' photo, taken four years later (see Figure 13), is displayed more like conventional mappa mundi – except without any cartographic grid.

Figure 14: Juan de Santa Cruz Pachacuti Yamqui Salcamaygua: Cosmology of the Inca, 1613

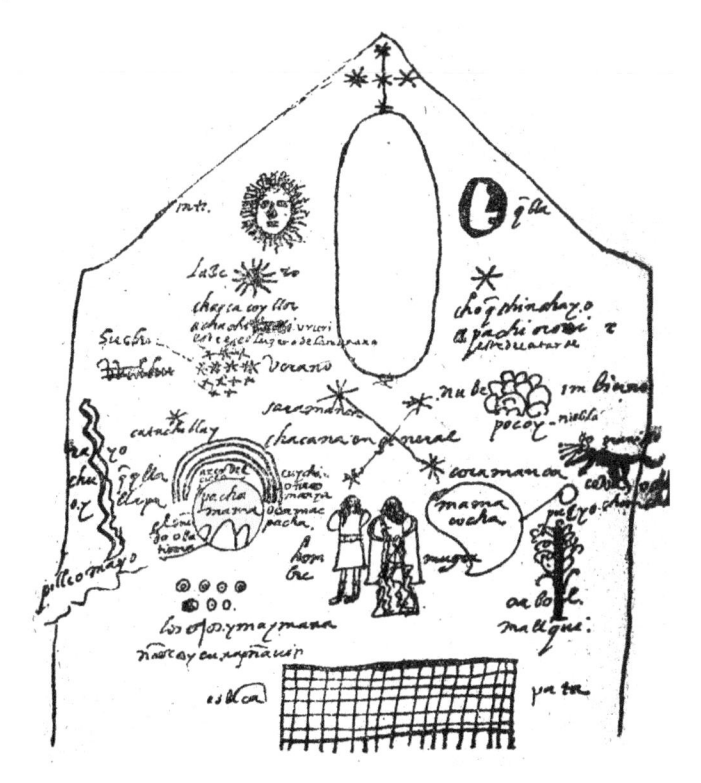

Source: Public domain

Juan de Santa Cruz Pachacuti Yamqui Salcamayhua, who recorded Figure 14, was an Indigenous Peruvian chronicler of the 15th century. At the time, Spanish culture sought to emphasise writing as its privileged system of recording, in contrast with the Quechua culture, which was based on oral and visual sign systems. Santa Cruz Pachacuti attempts to explain his native religion in the framework of the Christianity brought by the Spaniards, using visual images to go beyond language. Together with the works of El

Inca Garcilaso De La Vega and Felipe Guamán Poma De Ayala, his is one of the most important Indigenous sources of Inca society.

Figure 15: 5th Summit of the Non-Aligned Movement Logo, 1976

Source: Dinkel (2018)

The NAM is an organisation of nation-states that did not align with either of the two major Cold War power blocs (pro-Soviet, communist bloc or pro-American, capitalist bloc). Bringing together over 100 countries from across Asia, Africa, Latin America, the Caribbean and South Pacific regions, the NAM was founded as an organisation in Yugoslavia in 1956. Making up the majority of nations in the UN, the NAM's principle strategy was to democratise the UN and repurpose it as a tool for global justice (Prashad 2007). Figure 15 is thus less of a representation of the physical globe, as a representation of global politics and institutions at a moment when anti-imperialist and decolonial movements sought to fundamentally restructure global political and economic relations.

4

The Right to Subsist: Transnational Commons Against the Enclosure of Environments and Environmentalism

Captain Planet and the Planeteers was a US-based animated television series first aired in 1990. The story follows five young 'Planeteers' who have been tasked by Gaia, the spirit of the earth, to work together to protect the environment from pollution, deforestation and other environmental issues of the day. The five represent different regions of the world – Africa, the Soviet Union (later Eastern Europe), Asia, South America and North America. When they unite, combining their powers through rings bestowed by Gaia, they conjure up Captain Planet, a blue-skinned, green-haired avatar. Only called on in moments of crisis, Captain Planet always saves the day and then returns to the earth with the same message: 'The power is yours!' Meanwhile, opposing Captain Planet and the Planeteers are a cast of eco-villains with fantastic names: Hoggish Greedly, Verminous Skumm, Duke Nukem and, best of all, Dr Babs Blight, a mad scientist who conducts irresponsible experiments with science and technology (reminding us that industrial disasters such as Chernobyl, Bhopal and Three Mile Island were very much in recent memory). The shows always end with a short educational segment, usually about the need to be more conscientious about the environment but extending to topics like HIV/AIDS, drug abuse and even the Troubles in Northern Ireland.

In *Captain Planet* we see a simplified repetition of cultural and aesthetic tropes foundational to modern environmentalism as it was mainstreamed in the 1980s and 1990s. This mainstreaming was facilitated through the communication of key messages through the popular media, but also through the creation of new global institutions for collaboration around environmental goals, such as the IPCC established in 1988. The discourse of sustainable development became a defining way to obtain consensus on environmental issues in combination with cultural and political goals. Sustainable development was adopted as the overarching paradigm of the UN following the Brundtland Commission Report (1987), leading to the proliferation of global governance sub-institutions, such as community-based resource management initiatives. Once again, the definition of a single shared planet undergirded environmental strategies for unified action.

However, *Captain Planet* breaks away from earlier forms of environmental representation in important ways. One important difference is a sense of unbridled optimism, channelled through the youthful energy and always-victorious band of Planeteers. Chiming with the Brundtland Commission report released three years before the show first aired,[1] emphasis falls on rallying around sustainable development – including the new economic opportunities this offers – rather than radical social and political restructuring. By the end of the 1980s, a new world order was emerging, as can be seen in a second shift embodied by Captain Planet: the inclusion of a Planeteer from the Soviet Union (Linka). The Cold War that had dominated geopolitics and development since the 1960s was ending. The inclusion of Linka was slightly premature, but only by months, as the former communist bloc gave way under long-standing pressures. In this moment of Western optimism it was presumed that an end in ideological differences, symbolised by the fall of the Berlin Wall in 1989, would lead to a new global consensus around liberal democracy and a capitalist economic system.[2]

It is in this era that the 1992 UN Conference on Environment and Development took place in Rio de Janeiro, Brazil. Twenty years after the UN Stockholm Conference, the meeting, commonly known as the Earth Summit, brought together 108 heads of state, 2,400 representatives from NGOs and nearly 10,000 journalists over ten days (Sorensen 2012). The message of the conference was clear: new coordinated regimes of management and incentivisation were needed to avert an impending crisis. Informed by the Brundtland Commission's report, as well as the science emanating from the new IPCC, the Earth Summit gave rise to the key architecture of contemporary global environmental governance, including the framework conventions on climate change and biodiversity and the UN's Agenda21. The flagship term sustainable development captures the shift in this period toward models for addressing social and environmental problems as technical and governance issues, rather than as fundamentally political questions about power and ideology. This reframing of the problem also fostered a shift in governance away from state administration towards the alleged flexibility of market mechanisms and civil institutions.

Of course, not everyone was on board with these forms of environmental consensus-building and management. In this chapter, we trouble the growing consolidation of environmental governance narratives through the 1990s, turning to three movements and associated figures that asserted important disagreements *within* and *beyond* environmentalism. These movements and figures co-emerged with those already described, but modelled distinct agendas and aesthetics. Thus, we examine the emergence of the agrarian movement La Vía Campesina (LVC) in opposition to the organisation of global agriculture – a movement associated with the iconography of *Madre Tierra* (Mother Earth). Second, we document the movement of and for the Ogoni people, led by Ken Saro-Wiwa, as one of several environmental justice activists who generated international attention for their cause in this period. Third, we turn to the Zapatista movement in Mexico, seen by many as an iconic struggle against neoliberal globalisation and the defence of land-based territorial autonomy. Following the emergence of these movements and their constitutive claims allows us to argue that environmentalism became mainstreamed within new global institutions and policies only by sidelining (forcibly, in some cases) other perspectives, especially concerning resistance to structural adjustment policies.

The new enclosures

Glasnost,
    End of the Cold War,
    United Europe,
    We are the World,
    Save the Amazon Rain Forest ... these are typical phrases of the day. They suggest an age of historic openness, globalism, and breakdown of political and economic barriers. In the midst of this expansiveness, however, Midnight Notes poses the issue of the 'New Enclosures'.

                             Midnight Notes (1990: np; emphasis in original)

In 1990, the Midnight Notes Collective[3] exposed a glaring contradiction between optimistic claims of the 'end of history' and the violent expansion of capitalist frontiers via unprecedented waves of global enclosures. Such enclosures, as they termed them, proposed to dismantle various forms of 'commons' through privatisation and appropriation, while uprooting people from their livelihoods and homes, and degrading ecological environments. The collective underlined the economic violence imposed on countries across Latin America, Africa and South-East Asia via debt-induced neoliberal policies of privatisation, trade liberalisation and deregulation advanced by the Bretton Woods Institutions (the International Monetary Foundation and World Bank) and Structural Adjustment Programmes.[4]

Two years later, a special issue of *The Ecologist* magazine, written to coincide with the Earth Summit, makes a parallel argument (*The Ecologist* 1992). Entitled 'Whose common planet?', the collectively written text[5] also connects historic and contemporary processes of capitalist accumulation via the destruction of the commons. The authors extend this critique to the Earth Summit itself, and the emerging field of 'environmental management': 'the very same corporate and financial institutions pushing and benefiting from the "Washington Consensus" were confirmed at Rio as the key actors in the "battle to save the planet"' (*The Ecologist* 1992: 122). Recalling longer histories of peasant, fishers and Indigenous movements, they warn the environmental movement not to ignore the places and people struggling most intensely to defend their commons or regenerate commons regimes:

> [T]he demands from many grassroots groups around the world are not for more 'management' – a fashionable word at Rio – but for agrarian reform, local control over local resources, and power to veto developments and to run their own affairs. For them, the question is not *how* their environment should be managed – they have the experience of the past as their guide – but *who* will manage it and in *whose* interest. (*The Ecologist* 1992: 122; emphasis in original)

Both the Midnight Notes and *The Ecologist* make clear that sustainable development was becoming an agenda for *making markets work* for environmental protection, in contrast with questioning the structuring of the global economy around industrial capitalism and economic growth (Bäckstrand and Lövbrand 2006).[6] The new emphasis on managerial solutions proposed transforming the relationship between the state and the market to promote green investment, trade and innovation (Ziegler et al 2011).[7] This led to a proliferation of technologies for integrated pollution control, market-driven strategies to monetise environmental costs, and of models for flexible, decentralised policy making. New models like community-based management of forests and fisheries, and notions of 'stewardship of resources' among local communities also reflected this neoliberal turn, reallocating responsibility for meeting targets formerly held by the state and turning political questions into questions of governance and efficiency (McCarthy 2005; Bresnihan 2016b).[8] The technocratic focus of this agenda is almost silent on equity and poverty issues.

Events like the 1992 Earth Summit – a template for later environmental conferences – were part of the development of an environment discourse focused on participation and 'stakeholding' to build multilateral and polycentric cooperation in solving the environmental problem. However, what 'the problem' was became closed to question, with economic growth and environmental protection framed as compatible rather than at odds. Building on the sharp analysis of the Midnight Notes and the Editorial collective at *The Ecologist*, we see this as a key period in the struggle over the meaning and direction of environmental politics. Opposing the mainstreaming currents of environmentalism-as-management were land- and livelihood-based movements that objected to the continuing fall-out of resource-extractive and polluting development *and* the supposed environmental solutions put forward to remedy such problems. The goal of these commons-based struggles was 'not to win greater power for the market or the state, but to reinstate their communities as sources of social and political authority' (*The Ecologist* 1992: 122).

The emergence of La Vía Campesina as a response to neoliberal globalisation

A new transnational movement

The previous section sets the scene for the emergence of new agrarian movements dissatisfied with the ways that peasant autonomies and agricultural systems in the Global South[9] were being dismantled to make way for a new global industrialised regime. But mechanisms of food aid, from wheat subsidies offered to poorer countries after the Second World War, to new food security policies in the 1990s, were seen to structure in long-term dependency to already indebted economies. This dependency pushed countries in the Global South to convert their food production towards export-oriented crops, requiring them to sign up to long-term free trade agreements to receive any kind of assistance (Ziegler et al 2011; Carolan 2013). Over time, aid in the form of very cheap wheat or seeds from the US also transformed diets and economies, rendering whole areas like Central America commercially reliant on US production circuits and fluctuating prices (Edelman 2014). Yet the agrarian movements consolidating during the 1980s and 1990s sought not only to criticise such forms of enclosure but to model alternatives.

An important example of such a movement is LVC, a transnational movement of food producers and peasants set up during the early 1990s. Founding members linked the subsidisation of large-scale export crops with the large-scale dispossession of peasant farmers, as well as endemic forms of pollution (Desmarais 2007; McMichael 2014). In Latin America and elsewhere, the shared contours of resistance were tied to the introduction of the Green Revolution technologies, as explored in Chapter 2. Large-scale agricultural dispossession created impetus for bottom-up organising in subsequent decades, and an active revalorisation of agricultural knowledges framed as 'backward' or peripheral. In LVC, peasant, or *campesino,* farming is regarded as an important locus of agricultural innovation that needs systematic protection (Van der Ploeg 2012). This perspective resonates with that put forward by Mexican agronomists in the 1950s, as well as by the revolutionary African agronomist Amílcar Cabral in the 1960s, as covered in Chapters 1 and 2.

In 1996, LVC introduced their counter-concept, 'food sovereignty', to the UN World Food Summit, where definitions of food security were being approved. Food sovereignty was presented as a kind of 'agrarian citizenship' (Wittman 2011) with correlative rights for peasant farmers. Security is meaningless, it was argued, unless it protects the autonomy of peasant farming, and rights to *refuse* to open the best agricultural land to the vagaries of the global food market. Through dialogues and movement-building, LVC later grew into a 'peasant internationalism' comprising hundreds of thousands of peasant producers (Martínez-Torres and Rosset 2010).

Rather than 'solutions' that propose to save nature through markets, such movements want ecological modes for doing agriculture that make room for living well and the persistence of a multitude of traditions. Indeed, the knowledge production processes of movements like LVC have also been heralded as a potential model for democratic knowledge production (Pimbert 2006), because of the distinctive dialogic practice (*'dialogo de saberes'* [dialogue between knowledges]) that has shaped the movement since its inception (Rosset and Martínez-Torres 2012). LVC holds local, regional, national and transnational events instrumental in making manifestos and demands, and which have subsequently seen increasing invitations to mainstream environmental fora. Meanwhile, through practices of 'agroecological' knowledge exchange[10] producers share practices and ideas directly, experiment with new and traditional forms of agriculture, and consolidate practices that work in a range of experimental conditions (Rosset and Martínez-Torres 2012).

Returning to the broader theme of decolonising environmental politics in the book, the point here is that the agenda's force has emerged from an exchange not only between ideas, but between forms of knowledge that derive from diverse worldviews, cultural contexts and experiences of neoliberal globalisation (Borras 2010). These shared experiences are, above all, shared experiences of dispossession and exclusion from participation transformed into alternative claims and practices. As such, movements make claims not only on global food futures, but on the making of global environments and global environmentalism, which have been constituted without their interests or participation. The process of transforming this foreclosure is what some term 'commoning'.

Madre Tierra

Distinctive in this new moment of transnational alliance-building and knowledge exchange are new 'figures' of collective identification. What we mean by figures are individuals or tropes that derive from a particular location (or several) but are made to resonate with struggles in many other places. For example, since the mid-1990s, appeals to *Madre Tierra* [Mother Earth] have united differently situated activists and *campesino* (peasant) networks as part of political claims of food sovereignty and agrarian rights. Often imagined as a woman whose body is a mountain or the whole planet, *Madre Tierra* is commonly associated with the concerns of rural and Indigenous groups to protect ecological and cosmological systems marginalised by industrial food systems. As we explain later in this chapter, the creation of figures, slogans and ideas that could travel and resonate beyond the localised struggles of the Ogoni in the Niger Delta, and the Zapatistas in the Chiapas in Mexico, were also fundamental strategies in these movements.

The emergence of *Madre Tierra* onto the global stage reflects a political challenge to environmentalism as it was solidifying at this moment. Specifically, organising in the name of *Madre Tierra* was a way of arguing for the recognition of multiple ways for 'worlding the world', rather than replacing one form of one-worldism with another. When food activists and agrarian movements refer to *Madre Tierra* they are calling upon global publics to fight for a different kind of collective organising, where the right to be able to subsist is made something political.

There is something both ancient and novel in the use of a cosmological figure to make such claims. In Central America the dominant influence on the visual representation of *Madre Tierra* is the historical Maya civilisation that persists today in communities across Mesoamerica such as the Maya-Kaqchikel in the Guatemalan highlands, the Q'eqchi' in the Guatemalan and Belizean lowlands, and the Yucatec Maya in Southern Mexico and Belize. The Mayan 'cosmovision' became important to the translation of postcolonial ideas around and beyond Central America (via poetry and agroecology especially) because of its richly symbolic visual representations of ecological and planetary relationality. In particular, the notion of the primordial connection between *madre* [mother] and both maize and soil in Mayan thought has resonated across Central and Latin America, in the light of shared experiences of colonisation and the Green Revolution.[11] Like the concept of food sovereignty itself, then, *Madre Tierra* carries something of a strategic essentialism. It is a narrative-vehicle into which diverse experiences and cultural commitments are translated, without requiring that they compromise their 'autochthonous' singularities (Rosset and Martínez-Torres 2012).[12] While being heavily influenced through Maya image-concepts (or, in southern Latin America, Quechua image-concepts), *Madre Tierra* embodies a relational meshwork of Indigenous traditions, perpetually recreated in new contexts in dialogue with specific histories and conflicts (Millner 2021).

The intersection of concerns embodied in the iconography of *Madre Tierra* have consequently helped establish a shared set of witnessing practices that today connect countless agrarian and food-based social movements around the world, like LVC, who object to the social and ecological violence of the global food regime.[13] There are important critiques to bear in mind surrounding this figure. Ecofeminist analysis has consistently challenged the idea of 'womanhood' as a stable term, calling for a deconstruction of the qualities (for example, of emotional receptivity, docility and servility) that tend to be cemented into images of woman-*as*-nature (Warren 2000). On the other hand, reclaiming other, non-white, non-colonial imaginaries of planetary collectivity opens space for dialogue over what counts as 'natural' and to what degree this is 'universal'. Some emphasise that the making of such figures is already a 'queering' of nature because of the ways they redraw connections between gender, creativity, sexuality and relationships.[14] This deconstructive work is vital, because the openness of woman-nature figures is such that they can be appropriated even into white capitalist narratives. For example, in the 1990 ABC Earth Day Special, Bette Midler, as a hospitalised Mother Nature, is saved by capitalist consumers and good housekeepers – supporting continued capitalist growth *and* the notion that environmental problem-solving is just another domestic chore to be assigned to women (Alaimo 1994).

Box 6: Witnessing practices

Witnessing conventionally refers to first-hand experience of an event, presented to establish a truthful account, or, in religious terms, to attest to the nature or authority of God. However, an alternative genealogy of witnessing links Indigenous and postcolonial activism through the idea of speaking in the name of others not present. The notion of 'international witnessing' that informs humanitarian testimony of the atrocities of wars or violent regimes makes a similar connection: from the early 1970s aid workers and photojournalists linked with Médecins Sans Frontières broke with codes of neutrality previously associated with humanitarian assistance to 'witness' to the world (Ticktin 2011). Yet humanitarian testimony, like the activism of white environmentalists critiqued by Bruce Braun (2002) in his work on the British Colombian forests, is problematic because it speaks *on behalf of* silent others. Braun's front-page activist speaks 'in the name of wild things' but does not listen to the Indigenous groups who have inhabited the forest for centuries. Meanwhile, humanitarianism tends to render white activist citizens those who make political testimony, and powerless victims as 'others' in need of care.

In anti-colonial contexts witnessing is rather a truth claim made in the name of a collective subject. In Central and Latin America *testimonio* [witnessing] defines a literary genre that records the brutalities of colonial regimes. One of the most famous examples is the controversial testimony of Rigoberta Menchú (1984), a Maya-K'iche' Indigenous feminist who published accounts of her life through the Guatemalan Civil War to promote Indigenous rights. The account was controversial precisely because, it was argued, some parts of it were fictional, or came from the lives of other women. Yet *testimonio* follows different rules to other autobiographical literature:[15] it takes a collective as its subject and maintains fidelity to collective, rather than individual experiences, transgressing at the same time distinctions between public and intimate spheres (Beverley 1989). Indeed, when confronted by accusations that some elements had been fabricated, Menchú referred back to words from the introduction of her book: 'I'd like to stress that it's not only my life, it's also the testimony of my people' (1984: 1).

There are thus important tensions to take forward with figures like *Madre Tierra*. However, as we show in later chapters, much of this literary awareness and critique is present within the social movements that use such figures to connect and make concerns resonate. Indeed, *Madre Tierra* embodies a kind of 'bottom-up', transversal resonance that marks a response to, and a disagreement with, the kinds of universal 'we' imagined by North American environmental movements since the 1960s. As we explore in the third interlude, what we might call 'more-than-human witnessing' via tropes like *Madre Tierra* can transform speech conditions not only through their theatrical aspects, but by circumventing the 'territorial thoughts' of states and their politics of identity (Glissant 1989: 50; see also DeLoughrey and Handley 2011). These themes of translocal resonance, witnessing and subaltern environmental politics are further demonstrated in two contemporaneous movements: Ken Saro-Wiwa and the MOSOP in the Niger Delta, and the Zapatistas in Southwestern Mexico.

From the Niger Delta to Rossport: resource extraction, Indigenous rights and postcolonialism

The Niger Delta, with its face to the Atlantic, has been a site of (neo) colonial resource extraction for centuries. Long before oil deposits were discovered there in 1956,[16] the region was known for palm oil production, gaining the name 'Oil Rivers' in the 19th century – an eerie prophecy of what was to come. Before palm oil, slaves were traded here. Most recently, extraction has centred on 'black gold', or what Michael Watts describes as 'the fuel of our modern, turbo-charged, hydrocarbon capitalism and an item of exchange capable of achieving unimaginable wealth' (2009: 84). The double injustice is that while entire delta ecosystems are destroyed, and with that the livelihoods, culture and health of those peoples who inhabit them, the ensuing wealth generated has not accrued to the people most affected by oil extraction and production. Rather, it has been claimed by US and European petroleum companies, their shareholders, and a small but powerful political elite in Nigeria. In the late 1990s, 30 years after independence, populations in the oil-rich areas of the Niger Delta had failed to see their standard of living improve, or their ability to exercise control over their territories expand: formal decolonisation in 1960 did not bring any substantial improvements to the lives of the majority in the Niger Delta. This was particularly the case for the Ogoni people.[17]

Royal Dutch/Shell, in collaboration with the British government, began oil production in the Niger Delta in 1958. In a 15-year period from 1976 to 1991 there were reportedly 2,976 oil spills of about 2.1 million barrels of oil in Ogoniland, accounting for about 40 per cent of the total oil spills of the Royal Dutch/Shell company worldwide (FOEI 2019). The effect on the Ogoni has been devastating: along with the oil spills, gas flaring and waste discharge have polluted waters and degraded the alluvial soil, making livelihoods from agriculture and fishing no longer viable. The flaring of gas makes the air unbreathable, and the nights unsleepable – the artificial light from the flames stealing the darkness away (Nixon 2011).

In 1990, a movement began in defence of the rights of the Ogoni people, led by the charismatic Ken Saro-Wiwa, an Ogoniland native. Saro-Wiwa was a writer and lecturer who had previously worked as a government administrator in the Rivers State cabinet in 1968 – in the wake of the first Nigerian Civil War post-independence, the so-called Biafran War, which also related to control of strategic oil resources in the Niger Delta. In the 1970s, Saro-Wiwa helped rehabilitate Ogoniland within the re-established Federal Republic. However, 20 years later, inequalities between the north and south of the country had worsened. As the damage wrought by oil production accrued in land, soil and bodies, the Nigerian military government reduced revenues to oil-bearing areas, such as Ogoniland, from 50 per cent in the 1970s to 1.5 per cent in 1990 (Saro-Wiwa 2012 [1995]). In his memoir, *A month and a day* (2012 [1995]), written shortly before his execution in 1995, Ken Saro-Wiwa reflects on the lived experiences of these disparities as he travelled in a police car from Warri, an oil hub in Rivers State, to prison in Lagos in 1993:

> The state of the road irked me. It was one of my overriding concerns. Not the road itself, but the fact that in this rich, oil-bearing area, the roads should be so rickety, while in the north of Nigeria, in that arid part of the country, there were wide expressways constructed at great cost with the petrodollars which the delta belched forth. The injustice of it cried to the heavens! (Saro-Wiwa 2012 [1995]: 17)

In 1990, Saro-Wiwa helped write and disseminate the *Ogoni Bill of Rights*. The document traces the injustices experienced by the Ogoni to the history of British colonialism. The short document contrasts the US$30 billion raised for the Nigerian nation from oil mined on Ogoni territory, to the complete absence of representation in any federal institutions by Ogoni, the lack of pipe-borne water, electricity, healthcare and education services, shortages of food, land and job opportunities, environmental degradation and cultural demise, particularly of Ogoni languages. The Bill of Rights affirms the wish of the Ogoni people to remain part of the Federal Republic of Nigeria but with a greater degree of political autonomy, to include the control of cultural, economic and political rights; a fair proportion of revenues generated by oil production, and compensation and environmental remediation of past damage. As we see in what follows, these resemble the Zapatista demands in Mexico, whose vision of Indigenous sovereignty does not require the dissolving or takeover of the Mexican state, but its radical democratisation. Stemming out of the Bill of Rights campaign, a new organisation, MOSOP, was founded in late 1990.

Saro-Wiwa writes of two influential trips he took in early 1990 that influenced the formation of MOSOP and his own thinking on the Ogoni struggle. In the Soviet Union he witnessed first-hand the break-up of a multi-ethnic state. Other ethnic groups were struggling against the legacies of imperialism and political marginalisation, to gain recognition, rights and economic resources. Then, in the US, Saro-Wiwa witnessed the significance of organising to protect the environment. This was especially the case on his visits to Denver, Colorado, where he met with activists seeking to protect the old growth forests from the timber industry. Motivated by his own interest in trees, he recalls learning what could be achieved by environmental groups pressing their demands. This latter encounter is interesting as it disrupts the familiar hard line drawn between environmental protection campaigns in the US and environmental campaigns in places like the Niger Delta. Evidently, the contexts, protagonists and politics are miles apart, barely comparable, and yet there is space to find common cause and connection.[18] Saro-Wiwa explains in his memoir that it was these two encounters on his travels that determined the two major planks of MOSOP strategy: focusing on the marginalisation of the Ogoni people, and the devastation of the environment by oil companies.

Saro-Wiwa knew he had to reach the international community by whatever means possible,[19] but tapping into international networks and institutional spaces was not easy. When Saro-Wiwa contacted Greenpeace in 1991, he was told 'We don't work in Africa'. He tried Amnesty International and was asked: 'Is anyone dead? Is anyone in jail?' (Saro-Wiwa 2012 [1995]). This is where the blindspots of Western-centric but globally powerful discourses come to matter, demonstrating what is at stake in the aesthetics of environmental politics. The violence inflicted on Ogoniland was not *perceptible* within the confines of existing human rights and environmental discourses. New aesthetic resonances needed to be established to obtain international concern. Amnesty International only responded to the 'human rights' abuses that amounted to direct military killings or detention of people without a proper trial. Yet Saro-Wiwa insisted that the Ogoni were victims of an 'unconventional war', carried out on his people's ecologies and means of subsistence (Nixon 1996). Nixon (1996) aligns this example with other instances of 'slow violence' where the protracted timescale of the violence also keeps it from public perception. The fact that the Ogoni people had suffered these abuses for over 30 years (and more) is difficult to perceive *as violence* precisely because it does not manifest as a sudden catastrophe, but over a long period, in diverse arenas.[20]

Saro-Wiwa drew on all the contacts he had in the UK, where Shell was based and thus where public pressure might count for more, focusing on active, socially and environmentally concerned publics. Public pressure, including consumer-based pressure, were potentially powerful tools in the social justice campaigner's arsenal.[21] In 1992, Channel 4 made a documentary called *Heat of the moment*, and MOSOP were, eventually, able to get the support of Friends of the Earth UK and Survival International.[22] Saro-Wiwa managed to convince Greenpeace to send a team to witness the MOSOP protest march for the environment, held on 4 January 1993 in Ogoniland. This was to be remembered as Ogoni national liberation day. In Saro-Wiwa's speech, heard by hundreds of thousands, he connected the plight and struggle of the Ogoni to the Amerindians of North America and the Aborigines of Australia, and vowed that the Ogoni would not be destroyed like other Indigenous peoples.

As momentum grew around MOSOP, and with it international attention, the Nigerian government responded. In May 1994, Saro-Wiwa, who had been imprisoned multiple times over the previous two years, was abducted from his home and jailed along with other MOSOP leaders, in connection with the murder of four Ogoni leaders. Now Amnesty International acted, rallying international concern around Saro-Wiwa, advocate of non-violence, as a political prisoner of conscience. International pressure was limited, though. Meanwhile, under the rule of General Sani Abacha, Ogoniland was taken over by the Nigerian military, and the Ogoni people further terrorised. In October 1995 a military tribunal tried and convicted Saro-Wiwa of murder. Despite international condemnation, including calls on Shell to intervene, Saro-Wiwa and his eight co-defendants were hanged on 10 November 1995.

In the final weeks before his execution, Saro-Wiwa exchanged letters with Sister Majella McCarron, an Irish missionary nun, who had worked in Nigeria for 25 years. They first met in Lagos in the early 1990s because of her work with the 'Africa Europe Faith and Justice Network', a network of missionary groups in Africa that sought to raise awareness about the adverse effect European countries and companies were having on Africa (McCarron 2011).[23] Sister Majella recalls being drawn to the plight of the Ogoni people at the hands of Shell, but initially being confronted by a practical problem: Ogoniland was 500km from Lagos on roads that were poorly built (as Saro-Wiwa knew only too well). This distance from urban centres, national, regional and global nodes, is fundamental to making sense of Saro-Wiwa's activist work. It is also a fundamental theme in this chapter – recognising and appreciating the inventive ways social movements in otherwise remote places, lacking communications and transport infrastructure, translated their concerns and demands via new transnational networks and media: how concerns 'here' were able to travel 'there', and to what effect (Wenzel 2019). Sister Majella did manage to make the journey multiple times over the next several years, becoming close to MOSOP and, especially, Saro-Wiwa (see Figure 16).

In 2000, Sister Majella was back in Ireland watching the television. Images of a priest in a high-vis jacket blessing a gas well off the west coast of Ireland caught her attention (McCarron 2011). Shell had been given the rights to drill for gas in Irish waters, with plans to pipe the gas onshore and refine it in Belmullet – a remote, rural part of County Mayo. Sister Majella wanted to visit but quickly realised that, like Ogoni, Belmullet

Figure 16: Poem written by Ken Saro-Wiwa, 1995

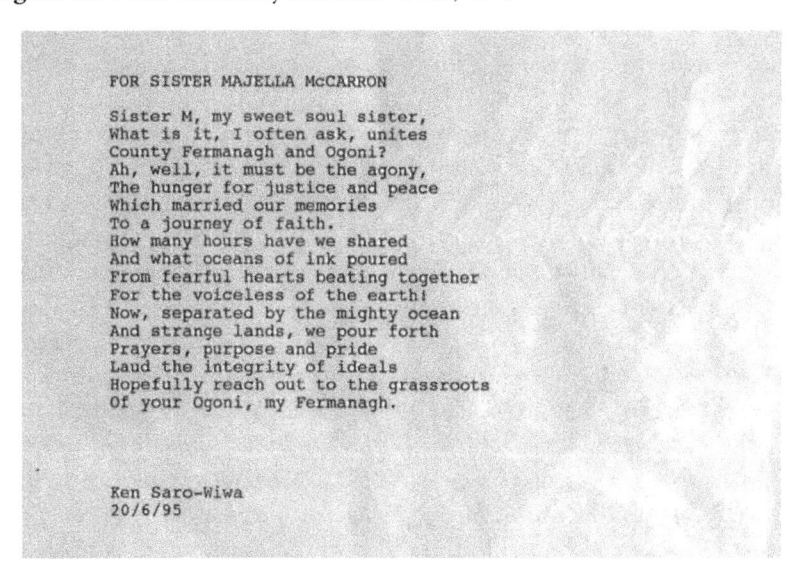

```
FOR SISTER MAJELLA McCARRON

Sister M, my sweet soul sister,
What is it, I often ask, unites
County Fermanagh and Ogoni?
Ah, well, it must be the agony,
The hunger for justice and peace
Which married our memories
To a journey of faith.
How many hours have we shared
And what oceans of ink poured
From fearful hearts beating together
For the voiceless of the earth!
Now, separated by the mighty ocean
And strange lands, we pour forth
Prayers, purpose and pride
Laud the integrity of ideals
Hopefully reach out to the grassroots
Of your Ogoni, my Fermanagh.


Ken Saro-Wiwa
20/6/95
```

Source: Ken Saro-Wiwa Archive at Maynooth University

was far away and poorly connected. Belmullet is a poor region of Ireland, relying on smallholder farming and fishing, livelihoods that rely on healthy environments. While miles apart geographically and historically, Sister Majella recognised the similarities with the place she had only recently returned home from.

Opposition to Shell's proposed refinery in the west of Ireland would remain fierce and protracted throughout the first decade of the 21st century; 'Shell to Sea' campaign based in Rossport, County Mayo, became established as the gravitational centre for the nascent alter-globalisation movement in Ireland. Almost immediately, strong links were forged with the Ogoni and their struggle against Shell, described by Saro-Wiwa as an 'ecological war'.[24] In 2005, at the height of the Shell to Sea campaign, a mural was unveiled at the site of the proposed gas refinery plant on 10 November by a four-year-old Ogoni boy, now resident in Ireland. The date marked the 10th anniversary of the executions of Ken Saro-Wiwa and his eight comrades by the Nigerian government. The mural features an image of Saro-Wiwa, right arm aloft, face lit with a smile, alongside lines from one of his poems translated into Irish: 'Dance your anger/and your joys/dance the guns to silence/dance, dance, dance'. After the ceremony, a party for Africans living in Ireland was hosted by Shell to Sea in the nearby village of Glenamoy, featuring African food and drink, a reggae DJ and African drumming and dance (Indymedia 2005). The mural and the connections between Rossport and the Niger Delta captures the unpredictable, endlessly hopeful power of connection between translocal sites of struggle that surfaced at the end of the 20th century (Gilmartin 2009).

Saro-Wiwa appreciated the power of words and story-telling and had seen that furthering the struggle for Ogoni political autonomy required the mobilisation of external influence and authority. This, in part, was why Saro-Wiwa took so quickly to Sister Majella. The missionaries were another remnant of the colonial past, but they were also a conduit to Europe; remnants of colonialism can be used to further decolonisation. Networks forming around similar but distinct struggles against state-sanctioned extraction and enclosure were able to take advantage of new institutional spaces for meeting and organising (for example, the Earth Summit in Rio), as well as the growing role of media and celebrity (for example, the Goldman Prize was established in 1990 to give recognition to 'ordinary individuals for outstanding grassroots environmental achievements'; Saro-Wiwa won the Goldman Prize in 1995 and Willie Corduff of the Shell to Sea campaign won it in 2007). Cultivating new aesthetic connections to carry these connections was vital to the new forms of solidarity that resulted. As we describe in the following section, the use of media, including the internet, and of iconic forms, would also be central to the ability of the Zapatistas to resonate far beyond the mountainous Chiapas region of Mexico.

Ya basta! Alter-globalisation and the Zapatista commons

> We do not ask for charity or gifts. We ask for the right to live in dignity, with equality and justice like our ancient parents and grandparents.
> Demands submitted by the Zapatistas during the February 1994 dialogue with the Mexican government (EZLN 1994: np)

On 1 January 1994, the Zapatistas declared war on the Mexican army, seizing the city of San Cristobal de las Casas and five towns in the mountainous region of south-east Mexico known as the Chiapas. 1 January 1994 was the day the North American Free Trade Agreement (NAFTA) came into force. In their communiqué to the national and international media, the Zapatistas explained that NAFTA would bring about the 'execution' of four million Indigenous Mexicans in Chiapas, the country's poorest province. But the rebels also made clear that NAFTA was only the latest in a long history of dispossession and violence experienced by the Indigenous peoples of Mexico. On that historic day, Commander Marcos, the charismatic figurehead of the Zapatistas, stood on top of the municipal presidential balcony in San Cristobal de las Casas, addressing the assembled crowd of approximately 400 people. Someone in the crowd asked: 'When did this start?' Marcos responds: 'Five hundred years ago! Was it not 500 years ago that our struggle against slavery started?' (EZLN 1994). '*Ya basta!*' or 'Enough is enough', became the rallying cry against 500 years of oppression, and the slogan for a generation of alter-globalisation activists around the world. The rebels called themselves Zapatistas, taking their name from Emiliano Zapata, the slain hero of the 1910 revolution who, along with a peasant army, fought for lands held by large landowners to be returned to Indigenous and peasant farmers. Land, commons[25] and the lived legacies of colonialism were at the heart of the Zapatista uprising.

Subcommandante Marcos became the elusive and quixotic (masked) face of the Zapatista movement. His biography is unclear, but he arrived in the Chiapas in 1984 with a band of fellow revolutionaries intent on educating, organising and leading an Indigenous, guerilla movement. When these young revolutionaries encountered the Indigenous world, they quickly found that they had to relearn everything from scratch. In the words of Marcos: 'Our square conception of the world and of revolution was badly dented in the confrontation with the indigenous realities of Chiapas' (Marcos 2011: 425). The new arrivals learnt that they were not the first group to imagine they were coming to 'educate' the peasants in the mountains. Bishop Don Samuel Ruiz Garcia and the church workers of the San Cristobal diocese had been intensely involved in Indigenous community life for over two decades. Bishop Ruiz had absorbed the radical currents of liberation theology during the

Second Vatican Council in Rome in the 1960s. Here we can see the less well recorded traffic of radical traditions moving back and forth across the Atlantic – the radical intellectual cultures of the 1960s in Europe, themselves informed by anti-imperialism in former colonies, rippling back out to the Lacandon Jungle.[26]

By the 1970s and 1980s, Mexican state repression in the Chiapas was intensifying and spreading, reinforcing the expansion of cattle ranching. Under these conditions, Marcos was able to initiate the establishment of a new kind of army, the EZLN, the Zapatista National Liberation Army, controlled not by guerrilla commanders but by the communities themselves, through clandestine councils and open assemblies. Marcos sought to fulfil revolutionary dreams and schemes, but these were charged by encounters with the Indigenous world in the Chiapas. It is the recognition, and working out, of difference that is important here – the practical and political question of how to craft something specific to the context, drawing on different positionalities and visions for the future. The evolution of Zapatismo is an inventive and experimental fusing of traditions and cultures that doesn't neatly fall into the categories of local/global, Indigenous/non-Indigenous, traditional/modern. From the beginning, the capacity of the Zapatistas to invert stereotypes and unsettle assumptions is precisely what confounded their opponents and created space for transnational encounters.

The southeast in two winds: A storm and a prophecy, written by Subcommandante Marcos (1992)[27] and first published in the weeks following the 1994 uprising, opens with a description of someone trying to get to the Chiapas. It is difficult because the area is remote from major cities and accessible by only three roads. And yet, Marcos writes, 'raw materials, thousands of millions of tons of them, flow to Mexican ports and railroads, air and truck transportation centers. From there they are sent to different parts of the world: The US, Canada, Holland, Germany, Italy, Japan, but with the same fate – to feed imperialism' (Marcos 1992: 1). Just as Ken Saro-Wiwa reflected on the poorly paved roads in the South of Nigeria, and the long history of colonial underdevelopment that produces these infrastructural deficits at the same time as the conditions for extraction, so does Marcos chronicle the resources that have been 'discovered', and drawn out of the south-east of Mexico – oil, coffee, timber, beef, honey, and the blood of those who inhabit the land. The metaphor of the vampire, now familiar in depictions of neoliberal capitalism, are evident from the beginning in the incendiary texts that emanate from Zapatismo.

The political aesthetics of the Zapatistas has centred on irony, humour and transgression – this quixotic element is not incidental, but part of a political project based on refusing the terms of recognition and inclusion offered by the Mexican state and the institutions of globalisation. The struggle for a different trajectory, the reopening of history, required different ways of expressing experience and power (Holloway 2005). For the Zapatistas, the use of the internet was part of a deliberate anachronism, a clash of temporalities that did not correspond to outside interpretations. They were not fighting against the 'modern' world to simply protect traditional ways of life. The Zapatistas refused temporal markers and assignations. From the beginning, their struggle was articulated not as the defence of the 'past' in opposition to the 'present' of neoliberal capitalism, but the expression of a *different* past, present and future that was no less 'contemporary' or 'modern' than the one being imposed by 'neoliberal global development' and 'Western progress' (Ceceña 2004). Their skill was to play against the grain, always positing hope in an alternative vision of the future that could draw equally on Indigenous stories, US popular culture and Shakespeare, for example.

As with MOSOP, international networks were vital to the protection and advance of the Zapatista movement. From the beginning, the attention of international activist communities meant that the Mexican government could not quickly and violently put down the Zapatista uprising (even if they had been able to). The Zapatistas knew that an isolated uprising in south-eastern Mexico could easily become another tragic but minor story in the history of state violence against rural, Indigenous populations. Thus, from their first communiqué in January 1994, they invited the international community 'to watch over and regulate our battles' (EZLN 1994). The summer after the uprising, they hosted a National Democratic Convention in the jungle; 6,000 people attended, mostly from Mexico. When the Mexican military sought to suppress the occupation, they found international observers. Unfortunately, this wasn't enough to stop all state violence. In 1997, the massacre of 44 men, women and children in a 'neutral zone' by paramilitary forces led to widespread outrage, and pressure on the Mexican government from abroad. In 130 cities, in 27 countries, on five continents, Zapatista supporters marched across Europe, Latin America, the US, Australia and Africa. In mid-February a delegation of 170 European politicians, intellectuals and human rights observers travelled to Chiapas on a fact-finding mission, and once again the military solution was averted. This is the power of global witnessing (see Box 6).

A key theme traversing many movements at this time was the demand for greater democratic participation and control over resources, as postcolonial states across the Global South buckled under the pressure of global economic forces facilitated by national elites. As in the claims of food sovereignty and MOSOP's efforts to protect Ogoni sovereignty and rights, the Zapatista demand was not for power over the state, but democratic rights within the state, including representation in federal elections. The Zapatistas called this the 'democratisation of democracy': the end to one-party rule, and electoral reform to prevent professional politicians from serving international financial institutions and corporations. This formulation unsettled the Mexican state, which had been quick to evoke the spectre of ethnic conflict in the wake of the Chiapas uprising. When it became clear that the Zapatistas were not an ethnic group seeking to secede from the Mexico state, the narrative switched, leading to claims that foreign forces were controlling Indigenous communities in their effort to undermine Mexico (Cleaver Jr. 1998). The Mexican state understood the uprising in the only terms it knew – either as a revolutionary move, or an ethnic move to secede from the state: both being an attack on the foundation of the state. But the Zapatistas fought for a different Mexico, rooted in the visions of revolutionary figures like Zapata, and symbols of a joint Mexican/Indigenous heritage. Unsurprisingly, this struck sympathetic chords in other parts of the world, and the Zapatistas inspired a new grammar of social movement organising and activism in the wake of the fall of the Soviet Union.

The Zapatistas are not often discussed alongside MOSOP or LVC, and yet they emerged at a similar time, in response to the same global political and economic forces, and to similar questions: how could they extend their experience and objections beyond their enforced remoteness? How to situate contemporary forms of dispossession within longer histories of colonial enclosure and environmental destruction? How to develop a place-based politics that could resonate transnationally? Naomi Klein (2001) compares the Zapatista uprising to previous Indigenous and environmental campaigns she had engaged with up to that point: until then, organising had been orientated around blockades, and the protection of land from the further incursion of resource extraction or polluting infrastructure. The Zapatistas were also involved in blockading: '*Ya basta!*' articulated a line in the sand that was not for crossing. But the Zapatistas did not leave the world behind. Their culturally and territorially specific struggle was articulated with and through figures, movements and campaigns from across the world. They established a key node in the emerging alter-globalisation movement, a transnational, transversal confluence of movements pitched together under common notions of dignity, justice and rights.[28] As with the figure of *Madre Tierra*, the Zapatistas cast themselves onto the global stage, contesting its occupation by the institutional figures of neoliberal capitalism and the master narratives of globalisation. 'We are all Marcos', the Zapatistas say, their iconic ski-mask becoming a means to reveal rather than hide, a mirror that reflects many different experiences of struggle against the inherited and ongoing injustices of imperialism and (neo)colonialism. At this time, the movements for the protection of the commons constructed a form of universalism (or what, in Chapter 4, we call a 'pluriversalism') beyond liberal democracy and Western capitalism, a universalism best articulated in the famous Zapatista slogan: 'a world where many worlds fit'.

Conclusion

The international experience and connections of LVC, MOSOP and the Zapatistas signal the emergence of new sites, movements and frameworks of objection to modern environmentalism throughout the 1980s and 1990s. As we have argued in this chapter, environmental concerns were being mainstreamed into public awareness during this period, with a new industry coalescing under the guise of environmental management that was seemingly more interested in buttressing Western models of capitalist economic development than in heeding the concerns of specific communities. At the same time, and in response to this failure to connect new ecological issues with colonial violence and fresh rounds of dispossession, Indigenous and local organisations were forging their own international networks and grammar, orientated around ideas of alter-globalisation and post-development. This sense of convergence was evident in some of the monumental environmental gatherings of the 1990s, culminating in the street protests of 40,000 people in Seattle in 1999. Such disagreements and aesthetic points of resonance and dissonance would also manifest at iconic summits during the 2000s as well.

Seattle would form a key reference point for political dissensus in years that followed, having been a site of erupting radical energies for many years. Yet visiting the city – home to transnational corporations like Microsoft and Starbucks, and where independent cinemas and clothing shops have become a brand of their own – this is difficult to imagine. Why is this relevant to this chapter about dissensual environmental movements responding to the enclosure of environments and environmentalism? Because the same patterns of gentrification and mainstreaming can also be identified in our examples. In some ways, agroecology has become mainstreamed – adopted into university programmes and business plans rather than known as a social movement; LVC is a huge social movement but has also been contained in important ways – now a minor speaker at the large forums it used to object to. Shell is still trucking oil, and the Zapatistas remain resolute but under constant threat of state violence (see Coda). Yet, there remain important currents of eventful political dissensus within and beyond such movements. As we have repeated in this book, we are not telling histories that are over, or lamenting what could have been, but tracing a different genealogy of environmental politics that continues to nourish and inform movements today. In concluding the chapter, we want to point to three important themes that signal that it is not the movements *per se*, but aspects of political aesthetics that render ecological struggles dynamic and a basis for organising against colonialism and capitalism.

First, the theme of icons or iconographies has recurred through this book and in our interludes. In this section we have seen how important 'figures' (like *Madre Tierra*, Subcommandante Marcos, Saro-Wiwa) are to the translation of a concern from one place to another, and to the amplification of that concern via the capture of new audiences. This was a time when 'transnational stories ricocheted here and there providing fodder for alternative dreams' (Tsing 2011: 215), a point that is helpful to understanding how localised, often ethnic or Indigenous struggles, like LVC, MOSOP and the Zapatistas, mobilised new kinds of global resonance to further their demands. In these cases, new vocabularies and symbols for intersectional movements within and beyond the Global South produced a different basis for collaboration than that offered by the UN or modern environmentalism, connecting common struggles while rejecting 'one-size-fits-all' solutions. What those who appeal to *Madre Tierra* share, for example, is a commitment to the way that countless ways of knowing 'nature', broadly defined, have been excluded within modern Eurocentric stages of politics. Witnessing is a powerful word to apply in such contexts because it reminds us that even as we feel and act alongside others, we *do* need to pay attention, careful attention, to each situation we arrive in, to know where we have arrived, and where from.

This question of witnessing and resonance links to our second point. In this book we argue that environmentalism loses its political purchase when it focuses on nonhuman land and ecologies *to the exclusion of* concerns with livelihoods, forms of labour and the violent histories of colonialism. Saro-Wiwa was part of a generation of environmental justice activists who could make such links across geographies and between the concerns of human rights, post-colonial and environmental activists. In the process, such examples revealed fresh possibilities for international movement-building.[29] These are vital points to remember, because the disruptive demands from below we have covered have frequently been folded into new regimes of governance that scale some aspects of agency and autonomy down to 'community', while leaving the 'natural' forces of uneven global trade and development in place (Li 2002).

The final theme we want to highlight is the coming to prominence of 'earth beings' – nonhuman persons like *Madre Tierra* that belong to more-than-human ontologies and make their appearance in environmental politics as part of new political claims. The notion of *Madre Tierra* was not new in the 1990s but a composite of older ideas: when she appears on the global stage she is making claims about what needs to be conserved and how. The notion of earth as *Madre* [Mother] arrives in Mesoamerican poetry, literature and activism in the same periods from a range of traditions, which, importantly, were newly entering global circulation (Millner 2021). To appeal to *Madre Tierra* in these contexts is to demand attention to the presence of human and nonhuman actors who have been exploited and silenced in the making of global capitalism – an act of denouncing wrongs on a vast scale – and also to claim, as the basis of collective action, a ground of lively and animating energies coded as 'resources' in capitalist forms of accumulation. In the next chapter we will develop this idea in more detail through the concept of political ontology, suggesting that, even though there is nothing new about earth beings, from the 2000s onward, something fresh emerges to challenge a sedimentation of environmental politics around management precisely through the staging of earth beings into new fields of collective concern.

Third Interlude: Witnessing in the Global Resonance Machine

Witnessing: communicating the truth of a lived reality, of a shared reality, in order to demand action. Speaking what can barely be spoken, because recognition is withheld. Seeking words, images, symbols and sounds that can hold together an experience that is being pulled apart, that requires new names.

> Rosa did not hide her anger as she got up to speak, raising a jar of soil above her head. As she spoke her words, she sprinkled the soil on the ground in front of her.
>
> "I speak in the name of all Guatemalans who are suffering hunger right now", she declared [translation Naomi's].
>
> More soil fell to the ground.
>
> "I speak in the name of those who are sick and do not have medicines. I speak in the name of families who don't have appropriate housing. I speak in the name of those who can't read and write. I speak in the name of the unemployed. I speak in the name of children, young people, adults, and old people. I speak in the name of the exploited. I speak in the name of the *Madre Tierra*, whom they exploit. ... In the past they obliged our grandmothers and grandfathers to pay tribute. They divided up our *pueblo* [village]. They stripped and killed our grandparents. And they sent us away to live in the mountains. This government does not represent us – its laws are an expression of the dominant class, protected by armed forces. We call for the creation of a Pluri-National state and the making of a National Assembly that represents our constitutive peoples. We call for a national strike. ... The *Madre Tierra* is the witness to our disfigurement, grandmother too wild to be tamed, and source of our future liberation. And so we will fight." (Fieldnotes, San Luis, Guatemala, June 2018)

This was a public meeting in San Luis, a municipality in the southeast of the Petén region of Guatemala. Rosa was speaking for the regional office of the Comité de Desarrollo Campesino (Campesino Committee of Development; CODECA),[1] a human rights and environmental rights organisation founded in 1992 to defend and promote the rights of Indigenous and rural workers. Attended by local and regional governors, representatives of international environmental organisations, the general public and national media, the meeting was part of a series of protests calling for the resignation of President Jimmy Morales and his deputies for blocking investigations into allegations of corruption.

In focus at this moment were the rising prices being charged by Energuate, a Guatemalan energy company now incorporated by Israeli-owned IC Power – that had repeatedly cut off those who could not pay the high prices. As well as demanding the nationalisation of electric power in Guatemala, rural workers and Indigenous groups were denouncing the environmental impacts of growing palm oil agribusinesses in the area, and calling for greater economic protections for peasant farmers in the face of growing cattle-ranching and incentivised large-scale export-oriented agriculture. Collective outrage centred on several recent kidnappings and assassinations of well-known activists involved in environmental protest.[2]

In this region, where a high proportion of the violent atrocities of the 1980s were committed, the Q'eqchi' Maya are still systematically excluded from state protections (Ybarra 2018).[3] As another speaker at the meeting put it, "these companies get what they want by making out we do not know how to manage our forests or our lives. This is the violence that we reject".

Rosa's testimony follows the Indigenous *testimonio* structure as she situates contemporary labour struggles in longer histories of colonial violence, speaking in the name of *Madre Tierra* to denounce the continued and systematic exclusion of the poor from basic services – but also to appeal to a collective subject that exceeds the forms of violence they have been subjected to. *Madre Tierra* is the collective subject of both this experience (she is 'witness to our disfigurement') and, of this future activism, as the ancestor who was not ultimately suppressed ('grandmother too wild to be tamed'). In this formulation *Madre Tierra* also exceeds the conventional 'mother' framing, being older and non-conforming, and she is also a source of collective memory against efforts to suppress and eliminate it.

As well as testifying against Energuate, Rosa testifies against the state government for their complicity in systematic corruption. She cites a series of cases, profiled by CODECA, where those contesting authorities were either killed, incarcerated or delivered from false accusations only after the intervention of international justice organisations. Lastly, and perhaps mostly interestingly, Rosa targets evangelical churches in the local area for upholding these authorities and for dissociating from the series of regional protests. "You are accomplices of the paramilitary and military", she goes on to say: "You speak in the name of Christ – but Christ at whose service? Christ did not defend the interests of the multinationals, but of the poor." Here Rosa refers implicitly to the movements of theology mentioned in the discussion in the previous chapter. The notion of 'witnessing' one's faith is transfigured here, through such speech acts, into a specific practice of denouncing colonial violence, while those participating in ongoing violence are portrayed as having 'false faith'.

Making things resonate: Joining together the bodies that have shared an experience so that they can act. Reaching beyond the bodies present so that others can hear and understand, and perhaps become part of a new collective. Finding an image or figure that stirs already so that a fresh movement might happen.

Problem: How to make things resonate in a way that means they are not presumed familiar already, which is to say: leaving space for this time to be different.

5

Earth Politics: Disagreement and Emergent Indigeneity in the So-Called Anthropocene

In June 2012, the Rio+20 UN Earth Summit – the third international conference on sustainable development hosted by the UN – opened with a short animated film entitled *Welcome to the Anthropocene* (Globaïa 2012). In just over three minutes, the film presents a history of the world since the industrial revolution, framed as the dawn of a new geological era, the Anthropocene. Progress, in the form of science, technology and industrial development, is depicted as emanating from England, bringing with it agricultural productivity, connectivity and population longevity. From 1950, graphs of environmental and socio-economic production appear, communicating a sharp acceleration in *rates* of transformation, as the planet is enrolled into a single consumption system. The film portrays this sweeping history through an unfolding three-dimensional computer model, with lines radiating out from Europe into a planetary network punctuated by bright nodes, which represent large cities. Layered atop of this cybernetic representation is a graph condensing dozens of different indexes of change in a single, upward trajectory: methane; forest loss; domesticated land; telephone connections; tourism; population.

The film was made by an NGO called Globaïa. Founded in 2009 by Félix Pharand-Deschênes, Globaïa's mission is the promotion of planetary awareness through 'the advancement of a science-based, transdisciplinary, emotion-driven and unified understanding of the major socio-ecological issues of our time' (Globaïa 2022: np). Besides producing the opening film for Rio+20, Globaïa have helped design significant environmental iconography, collaborating, for example, with the Stockholm Resilience Centre on their Planetary Boundaries project, and David Attenborough for his latest Netflix documentary (Globaïa 2022). Globaïa's film inherits much from the cultural and aesthetic tradition we have associated with modern environmentalism in

... We witness a single planet floating in space (our spaceship home); ... conveying a view from nowhere in their call to protect it; images ... ingly unpeopled environments as that which is under threat; emphasis on the urgency of working together in unified action to avert further crisis. As we have also argued throughout the book, such aesthetic devices conspire to sidestep a more critical reckoning with root causes and extent. They keep invisible the multiple sites and movements of resistance that have challenged the smooth unfolding of colonial modernity – what Stefania Barca (2020) has recently called the 'forces of reproduction'.[1] Against this grain, Marxian, decolonial and Indigenous scholars and activists call for 'a critical project that understands that the ecocidal logics that now govern our world are not inevitable or "human nature", but are the result of a series of decisions that have their origins and reverberations in colonisation' (Davis and Todd 2017: 763; see also Moore 2016). This chapter follows these alternative calls to the point where they reframe and recalibrate environmental concerns.

The Globaïa film is not the focus of our critique: instead we target the new aesthetic and techno-scientific framework that unfolds across the first decade of the 21st century.[2] The idea of the 'Anthropocene' is a key element of this framework, which emerges – with great accolades – from new ways of mapping and narrating the geo-social histories of global environmental change (Bonneuil and Fressoz 2016). Combining representations of complexity with the urgency of planetary extinction, this idea functions to transcend sensibilities of historical accountability, responsibility and justice, even as it seems to acknowledge the importance of changing trajectories of human production. This has given rise to a form of planetary politics dominated by geosciences and computer models, large-scale datasets and remote sensing, algorithms and automation (Gabrys 2016; Yusoff 2017).

Producing the world at such scales carries important implications for who can respond and how. Earth systems and climate models are complex, statistical technologies that are not possible without countless monitoring devices, scientific standards and institutions, and computer processing capacity (Edwards 2010). These models are not disembodied or abstract: they are produced through myriad forms of labour by scientists and technicians, and yet the way they are presented creates the illusion that they are. They serve to further the abstraction – or better, the semiotic openness – borne by images like Earthrise (see Chapter 3), used to guide calls to environmental action. We are thus invited to appreciate an undifferentiated humanity living on an undifferentiated earth as part of an appeal to unified action. Forms of technical expertise are presented as vital to solve the 'problem', due to the problem's scale and complexity, creating the conditions for technological and market solutions to dominate.[3]

Yet the emergence of this global technovision coincided with the consolidation of alternative planetary visions. In the same period (2010–2012) other seminal gatherings of world leaders took place in the name of 'earth politics', reimagining global cooperation from below. By earth politics, we mean ways of articulating collective action in relation to biodiversity loss, extinction and climate crisis that are planetary, earthly and/or geological – like the Anthropocene, they engage with the matter and life of the planet as part of making calls to action. Yet, unlike the Anthropocene, these modes of articulation break away from the one-world vision we have critiqued, seeking to pluralise it. Indeed, it is part of the premise of earth politics that excluding Indigenous and non-Eurocentric modes of planetary inhabitation is what got us into this mess (Simpson 2017). In this chapter we aim to contrast these forms of collective politics with the aesthetics of the Anthropocene.

We lay out the context for the appearance of these alternative forms of earth politics by documenting the planetary debates in play at a 2010 summit in Cochabamba in Bolivia. We then mobilise key concepts defined through these events to unpack two further examples of commoning between Indigenous modes of doing and non-Indigenous communities: Salvadoran permaculture practices, and Indigenous resistance at Standing Rock. These examples involve vastly different historical, social, geological and political formations and ecological practices but also capture aspects of what we are calling in this chapter earth politics. Specifically, they reveal how the Eurocentric, humanistic models of politics exclude non- and alter-modern forms of knowledge and knowing. The presencing of what are termed 'earth-beings' across these cases forms an important way in which this disagreement is made known. Following these examples also helps us show how earth politics models alternative forms of collective politics based not in universality, but pluriversality. Political disagreement at the heart of modern environmentalism is pushed further here than in previous chapters: rather than disrupting accounts of *which humans* are present and can appear, or which contexts are considered relevant, here persons who *are not persons* in the modern order of things make an appearance, making ontology itself a matter of dispute.

Provincialising the Anthropocene: the People's Earth Summit

The World People's Conference on Climate Change (WPCCC) took place from 19 to 22 April 2010 in Cochabamba, Bolivia. Contextualised by frustrated responses to 15 years of international climate negotiations led by the UN-organised Framework Convention on Climate Change, the WPCCC was billed as a counterpoint to the 2009 'Copenhagen Accord' – the widely hailed 'non-agreement' resulting from the COP-15 Summit.[4] Called by Bolivia's first Indigenous president, Evo Morales, and organised in conjunction with the emerging counter-bloc of 'progressive' governments in Latin America, 'ALBA' – the Bolivarian Alliance for the Peoples of our Americas[5] – the summit was a consolidation of a counter-movement created to devise alternative strategies to the (neo)liberalisation of labour markets, tax reforms, privatisation and austerity measures during the 1990s. Bolivia had been on the frontline of neoliberalising processes in the wake of the Washington Consensus in the mid-1990s (Held 2005) and the neo-Keynesian consensus which followed (Brenner et al 2010). However, according to the Bolivian government, of the more than 35,000 participants who attended the WPCCC, almost 10,000 travelled from abroad, with some 140 countries represented (Lindisfarne 2010). While the substantial majority were from Latin America, there was attendance from North America and some from Europe, Asia and Africa. As such, the event was characterised by a convergence of heterogeneous networks from the Global South and North attempting to imagine the world beyond neoliberalism – like the World Social Forum and the alter-globalisation movement in other geographic contexts that also took off in the 1990s (see de Sousa Santos 2005; Roelvink 2010; see Chapter 4), as well as the anti-imperialist Third World alliances and NAM in motion from the 1950s (Prashad 2007; see Chapter 2).

While building on decades of experimentation with fresh iconographies to dissent with neoliberalising global economies, the Cochabamba summit also marked the rearticulation of the *Madre Tierra* figure, discussed in the previous chapter, with new significance. From the 1990s, *Madre Tierra* was established as a unifying ground for collective politics for agrarian social movements of quite distinct political and social backgrounds. The Mayan cosmology continued to form an important influence on aesthetic representation in these campaigns, although as such movements multiplied, many Indigenous and rural groups came to share identification in icons like *Madre Tierra* in new forms. The overlapping figure of *Pachamama*, an earth mother in the Quechua cosmology, took on a wide resonance among social movements in the following decade, this time focusing on what it would mean to make Nature a subject of legal rights (De Angelis 2011). At the Bolivian gathering, *Pachamama* therefore united multiple Indigenous, *campesino* and other groups present at the gathering sharing a sense of dispossession over extraction and land rights issues – but also seeking a different means to resolve them.

As an image of planetary connectedness and relationality, *Pachamama* seems to reflect the planetary unity that underpins modern environmentalism. Yet, rather than making the Quechua cosmology the 'right' one, *Pachamama* is an invitation to look again and see that many 'worlds' exist on the earth, opening up different ways of knowing. This also makes the connection that environmental violence is crucially bound up with violence to entire knowledge systems. The Globaïa film repeats such violence when it proposes technical solutions that obscure how technical solutions have been part of the problem. The text of the People's Agreement, on the other hand, makes clear that global scientific knowledge and frameworks – the startling figures documenting biodiversity loss or carbon emissions, for example – cannot be tackled without wider-reaching processes of economic, cultural and ecological reparation.

Scholars in anthropology, human geography and political ecology suggest that this fresh emphasis on the politics of knowledge is central, not peripheral, to a decolonial environmental politics (Escobar 1998; Rose 2004). From the 1990s moving into the 2000s and to the present, such concerns have been increasingly framed in *ontological* terms, which is to say, in terms of the accounts of the world and its liveliness that animate diverse systems of knowledge and practice around the world (de la Cadena 2010). When Indigenous ontologies are taken seriously, it becomes clear that not only human subjects, but living and non-living matter can be regarded active participants in disputes over land and environments.

While this 'ontological turn' emerged over the last couple of decades in academic thought, the ethic and practice of relating to the nonhuman world as creative and sacred has, as we have signalled, a much longer history (see Todd 2016). In the debates over the rights of nature that animated protest in Bolivia and Ecuador in the decades preceding the Cochabamba event (see Ari 2014), and in new aesthetic traditions such as afro-futurism and Indigenous futurism, emphasis falls on addressing the denial of *worlds*, historical and actual, that colonial appropriation has enabled (Simpson 2017). Through the perspective of decolonial environmental politics, this violence is perpetuated in certain models of environmental governance.

We thus arrive at a central idea articulated in contemporary decolonial environmental politics – the politics of *incommensurability*. Incommensurability is the idea that different ways of worlding the world cannot be reconciled – and that is where international collaboration needs to begin. Building on this idea, decolonial anthropologist Marisol de la Cadena (2010) builds on the notion of the pluriverse, also mobilised by diverse social movements, as an alternative to the assumption that finding unity/universality is the way to solve environmental and political problems. The pluriverse is a direct affront to projects of building universal bases for rights and law in the project of modernity. While pluriversal scholars do not reject the ethos of building bridges between diverse cultures, nor of constructing shared principles of rights and justice, there is an objection at the heart of this concept. It is asserted that what has been called universal is highly specific and local to Anglo-European forms of knowledge, violently imposed as the norms of civilisation through the colonial project. Such forms class as primitive or 'only cultural' central elements of other cosmologies, such as what de la Cadena calls 'earth beings' – sentient beings that 'do not inhabit, but that *are* mountains, rivers, lagoons, and other visible marks of the landscape' (Micarelli and Verran 2018: 123). Pluriversal commitments consequently open out onto multiple designs for living well in the world (Escobar 2018). Such visions begin with the idea that collaboration towards the construction of just and ecologically sustainable practices must begin with disagreement over the givens of the known world, and acceptance that there is no one blueprint for civilisation. This must take place *before* the elaboration of any 'global solutions'.

Signalling Jacques Rancière's notion that politics is not made up of power relations between given actors in a common world, but of relations between socio-natural 'worlds',[6] de la Cadena (2010) emphasises environmental struggles in the Global South as a negotiation of the relationships between non-commensurable forms of worlding. She is interested in how earth beings – that have vital existence in one world but are not recognised in others, above all, in the 'modern' one – break onto political stages that do not yet recognise them as political entities to transform such relationships. Claiming reality and meaning in spheres structured according to rules that refuse them is what renders 'earth beings' as contentious objects (de la Cadena 2010: 342), and promises to 'reshuffle' hegemonic principles that have structured environmental governance for centuries (de la Cadena 2010: 343).[7]

More recent work on Indigenous political ontologies draws on similar arguments to contend that what is at stake in resource extraction conflicts is not one common nature fought over by antagonistic social forces, but fundamentally divergent relations to nature. Beyond the Latin American context, this marks a challenge to the universalist assumptions of Eurocentric political ontology, which presume one common natural world, understood via diverse social worlds. For Escobar (2010: 3) and de la Cadena (2010), the inclusion of earth-beings into national constitutions like Bolivia's therefore signals a potential move towards a more plural set of modernities, where diverse ways of knowing and valuing are taken seriously.[8] Summits like Cochabamba expand such ideas to demand future agreements on dramatically different terms.

There are two steps to the enactment of the pluriverse in the account set out by de la Cadena (2010: 361): we have, first, to recognise that the world is more than one socio-natural formation; and, second, to interconnect such plurality *without making the diverse worlds commensurable*. In the examples we unpack in this chapter, we trace and map out these two steps, contrasting the forms of association and earth politics that emerge from practice with the images of environmentalism portrayed by Globaïa. Each example explores aspects of what de la Cadena calls (2010) an 'emergent indigeneity' – which is not a new mode of being Indigenous, but an 'insurgence of indigenous forces and practices with the capacity to significantly disrupt prevalent political formations'. In the following sections we therefore explore how two contexts shaped both by earth politics and by internal differences are articulated in an overlapping, but non-identical, way. In both contexts Indigenous actors come together with non-Indigenous actors in new ways, or make Indigeneity part of a politics and iconography that brings together a community who do not identify as Indigenous *per se* (and see Box 7 for a further example). We find these cases interesting because sometimes there is a tendency to assume that earth politics is only relevant to the presencing of Indigenous cosmologies, in Indigenous-only contexts. In the remainder of the chapter we explore how earth politics helps us rethink modern environmentalism in more expansive ways.

Box 7: *Wild Relatives* (2017) by Jumana Manna: worlding seed conservation from below

In 2015, the International Centre for Agricultural Research in Dry Areas (ICARDA) became the first agency to withdraw a deposit of seeds from the Svalbard Global Seed Vault. The seed vault was established in 2008 as a 'Noah's ark' for global biodiversity, collecting and storing hundreds of thousands of seed varieties in the deep, former coal mines of Svalbard, an island in the Arctic Ocean. The seed vault is described as a 'long-term seed storage facility, built to stand the test of time – and the challenge of natural or man-made disasters'.

Originally based in Aleppo, ICARDA withdrew its seed deposits (originally deposited in 2008) at this time due to the ongoing civil war in Syria. The Centre had been forced to evacuate to Lebanon, leaving behind equipment and seeds. Later, in 2017, representatives from ICARDA were able to return to Svalbard with duplicates of the seeds that had been 're-woken' from their frozen state, planted and cultivated in Lebanon's Bekaa valley, and carefully dried and transported back to the Svalbard seed vault.

Jumana Manna's 2018 film, *Wild Relatives*, takes this seed transaction as the departure point for her beautiful film exploring the geopolitics, histories and cultures of seed conservation in the Middle East since the mid-20th century. Looming large in the narrative is the Green Revolution which sought to transform agricultural production around the world via high-yield seed biotechnologies, chemical pesticides and irrigation infrastructures. The key institutional nodes responsible for globalising these 'progressive' agricultural practices were the regional agricultural research centres, such as ICARDA, set up in the 1970s.

ICARDA was therefore initially part of the modernising mission to replace traditional, local seed-saving practices, rationalise land-holdings and make agriculture more productive across the world's 'dry areas'. These technologies and practices drew farmers into a system of dependence: on the state, the companies who sold the seed, and the energy and chemical inputs they required as production was scaled for expanding markets. The 'landraces' that farmers derived from 'wild relatives' – species that are untended by humans and continue to evolve in the wild developing traits such as drought tolerance – declined as the practices and knowledge required to produce such *in situ* seed diversity were replaced.

Jumping to the present, the film connects this historical rupture in land and seed relations to present-day events – the war in Syria that was preceded by historical droughts; the market reforms that make it nearly impossible for Lebanese farmers to make a living from the land; and the transaction with the Svalbard seed vault.

The seed vault captures the character and assumptions of the 'good Anthropocene'. It is premised on the idea that we can 'fix' planetary problems, such as biodiversity loss, through new, technological solutions implemented with urgency and at scale. In the film, this is illustrated through a series of conversions: the former coal mines transition from mineral extraction to seed storage, the supply chains and infrastructures of global commodity markets convert to transfer seed genetics, and the agricultural centres that once pushed for seed specialisation shift to the science of biodiversity cultivation. And yet, as the film captures, the forms and relations of power invested in these industrial landscapes, infrastructures, institutions and discourses are stubbornly unchanging. Despite involving good intentions, enterprises such as seed vaults are exemplary of a centuries-long colonial project that holds out 'a promise of techno-fixes premised upon and extending the logic of imperial botanical taxonomy' (Sheikh 2018: 224).

Imagining planetary seed conservation outside the confines of the global seed vault project, Jumana Manna also brings us into the world of Walid, a Syrian refugee who has taken up residence in the Bekaa Valley, near the new ICARDA station. Walid, we learn, had previously lived not far from ICARDA's Aleppo station. Walid, however, practises permaculture and makes natural pesticides. He describes the soil as having 'a world of its own', and stays connected to farmers back in Syria to share knowledge about farming and seed saving. Unlike the global, 'public' seed vault, reliant on inherited institutional and infrastructural blindspots, Walid tends to a plot of land using organic methods, preserving 'landraces' in glass jars and paper bags in a makeshift seed library.

Walid's seed relations offer us different remnants of the past that endure today, a commons beyond the private spheres of agribusiness and the 'public' spheres of state management and conservation (Sheikh 2018). These pockets of alternative practice may seem minor and inconsequential alongside the scale and might of the global seed vault, but their endurance through, or despite, the Green Revolution, neoliberal reform, and war, carries, we think, something hopeful.

Earth politics in Salvadoran permaculture practices

While earth politics in Cochabamba focused on the rights of nature and the place of earth beings in constitutional rights, elsewhere in the world at this time, collectives were also asking important questions about planetary politics at other scales, including the microbial. For just as concepts like the Anthropocene rippled through transnational networks, offering ways of making solutions, so *Madre Tierra* also morphed through debates about Indigenous witnessing, food justice practices, the rights of rivers and soil repair. We do not mean to suggest here that *Madre Tierra*/Anthropocene were becoming a new binary concept, nor that the former was enacted everywhere as a deliberate affront to the latter. Rather, the 2010 moment was an important time for thinking about planetary futures. And, for those disagreeing with the large-scale technical ways that modern environmentalism was taking shape, *Madre Tierra, Pachamama*, earth politics and pluriversal experimentation formed an important conduit for thinking collective action on the one hand, and everyday ecological practices on the other. In other words, this was becoming a new site of creative experimentation for *alternatives* to environmentalism, where the unquestioned 'one planet' at the heart of modern environmentalism was being rethought via relational, multiple earth-being. In this section we turn to what this implies for everyday soil repair practices.

Soil repair is necessary because the world's soils are highly degraded – about a third of all soils are classified this way in a recent report to the UN, while the world's topsoil could be fully unproductive within 60 years (Maximilian et al 2019). In many cases, this degradation can be closely linked with the rapid uptake of Green Revolution technologies from the 1960s onwards. El Salvador, where the Salvadoran Permaculture Movement – formerly the Permaculture Institute of El Salvador – is based, was a key site where techniques developed in Mexico were expanded (see Chapter 2). As in Mexico, El Salvador has also been a site for experimental approaches to ecology and knowledge conservation, such as agroecology, although this only really expanded here at the end of the Salvadoran Civil War (1979–1992). The war suspended many forms of agriculture, but also enabled many of the forms of popular organising upon which agroecology would become based.

Permaculture does not share quite the same roots as agroecology, being an ecological approach to holistic landscape design and food-growing that has developed in Australia and the Global North – but has overlapped with it considerably in Central America.[9] Permaculture principles centre on learning to observe and imitate biophysical processes, such that systems created to meet human needs (for example, food production), complement pre-existing ecological dynamisms and reduce the need for additional inputs of energy, materials or water. In Central America these principles have been found to resonate considerably with incipient principles for *Madre Tierra* praxis derived from remembered and revived Mayan and other Indigenous agricultural practices. While permaculture flourished more in Australia and the Global North initially – and in some cases has been seen or even practised as a kind of colonialist eco-tourism – in El Salvador it has continued to overlap with the networks and practices of agroecology. As Puig de la Bellacasa (2010) suggests, also in relation to permaculture, it is a biological principle based in a practical ethos that sets new horizons for care for species life, and is coherent with the alternative earth politics claims constituted through *Madre Tierra*.

In parts of rural El Salvador, agroecology came to mix with permaculture practices from 2000 onwards. For example, permaculture arrived in Suchitoto, a mountainous municipality of El Salvador, via a self-declared permaculture 'missionary' returning from Australia after a period of exile, sponsored by Christian foundations who saw continuity between the principles of liberation theology and agroecology. Liberation theology, as outlined earlier in this book, was an important way that participatory processes spread throughout Central America. Also, in contexts like El Salvador, where agroecological collectives have tended to form around existing political loyalties and commitments, permaculture enabled new kinds of collectives, with women and young people taking on more of a leadership role. Through the figure of *Madre Tierra*, Salvadoran permaculture soil repair is imagined as a kind of ecological and decolonial healing.

In El Salvador, experimental soil repair takes place through composting practices and the creation of biofertilisers – recipes that use local organic ingredients to boost soil health (Millner 2016). There are several words used for compost on a Salvadoran farm: *abono* is decomposed vegetable matter, while *fertilizante* denotes chemical fertilisers. Bokashi, on the other hand, is active, fermented organic matter. The term entered the lexicon via Costa Rica in agroecological exchanges, where farmers learned to mix animal and vegetable wastes with other substrates (for example, eggshells, ash, soil and coffee grounds), and to add to these a blend of microorganisms sourced from mountainous areas along with active ingredients such as molasses and yeast to help break down these wastes. This produced a special kind of fermented compost that became important to *campesinos* involved in agroecology and permaculture, because it could be made for free from local ingredients and had an observable impact on yield and crop quality.

Bokashi itself is adapted from a Japanese agroecological practice that travelled to Central America via Costa Rica through the *campesino-a-campesino* (farmer-to-farmer) movement that translated agroecology practices across Central America during the 1980s and 1990s (Holt-Giménez 2006).[10] In the Salvadoran process this recipe had been adapted to involve collecting naturally occurring mountain microorganisms, to activate the process. As agroecological farmers have adapted and shared bokashi, so its recipes have multiplied, reflecting local availability but also the variability of traditional practices. Usually, the process involves arranging, in a covered area, solid materials (such as bran, rice husk, soil, manure, charcoal and ash) into a pile, together with a source of easily assimilated carbon (for example, molasses, honey or sugar) and a mixture of microorganisms dissolved in water (Quiroz and Céspedes 2019). Yoghurt or yeast is used to multiply the effectiveness of the microorganisms. The pile is then covered with plastic or sacks and is turned several times a day for a period of 7–21 days (the recommendation varies). Although there are few scientific papers on its use, those that exist suggest that it works by allowing the slow release of nutrients, in contrast with chemical fertilisers that can leach quickly through rainfall and gas release.

Bokashi is not an Indigenous concept, then, but part of a repertoire of lay scientific practices developed experimentally in the context of the agroecological networks developing since the 1940s in Central America and elsewhere. Indeed, one reason for the lack of systematic research to date is that agroecology networks have flourished outside, and in resistance to, institutionalised soil science research communities – as we have shown in earlier chapters. Bokashi is an example of a technique that involves combining different kinds of knowledge and input, in contrast to the emphasis in mainstream biophysical scientific approaches on isolating single processes and effects. These translocal networks of alternative soil repair practice also represent as a kind of *Madre Tierra* praxis – a way to embody everyday interactions with a degraded earth also understood as a kind of sacred ground or even person/family of persons. What is interesting in this example is that earth politics can be seen to ground not only public forms of international witnessing, but intimate involvement with 'planetary' processes including soil formation, which, in turn, come to mediate collective processes of grief and healing.

Within the counter-narratives animating soil repair practices in agroecology and permaculture, soil becomes a focus for collective memory and recuperation. The violence associated with the civil war and genocidal warfare on Indigenous groups and *campesinos* took place in such fields, which also stand as a material record of ecological degradation. Soil is thus a kind of 'archive' of more-than-human memory, embodying elements of future agriculture and survival. In agroecological and permaculture movements, the erosion of traditional and Indigenous farming practices is also associated both with the impoverishment of soils (implying the loss of food security) *and* the destruction of livelihoods and associated cultural know-how. It is understandable, then, that soil repair practices are often imagined within such practices as a kind of decolonial healing process (see Millner 2017). The idea here is that repairing the damage incurred on the fabric of subsistence agricultural systems (soil) can take place through the rehabilitation of marginalised traditional and Indigenous knowledge systems on the one hand, and the rebuilding of economic autonomies on the other. In the process, new innovations are celebrated and what counts as 'traditional' is expanded to include practices developed by agroecological communities elsewhere, such as bokashi, which share the same textures of repair and critique. Soil repair, here, is not just a question of restoring traditional practice, but developing a form of lay experimental science that rejects the notion of the field as a flat space without histories, or a stage for modern geopolitics.

While experimentation in action-oriented pedagogies of agroecology and permaculture is never about uncritically restoring past models of practice, Indigenous techniques and motifs do become revalorised in this process. In particular, the notion of the primordial connection between *madre*, maize and soil resonates across Central America, especially in the light of shared experiences of colonisation. As we unpacked in Chapter 4, in Mayan portrayals of soil life, both what has been lost (land, cultural practices, soil fertility), and what is to be restored, are joined together, centring on terminology of *madre*, which links in Mayan language to mother, maize and soil (see note 11 in Chapter 4 for a review of Yucatec Maya understandings of soil). In everyday storytelling of the Green Revolution, as through the bokashi practice, soil is imagined as a living body rendered addicted by pesticides, that can be healed and restored through experimental biotechnologies and practices of care.

The creative energy of such praxis manifests, for example, in the performance of sociodramas, which farmers use to communicate agroecological practices from one community to another. Through permaculture workshops groups create their own interpretations of the Green Revolution, based on oral histories and new learning. In one performance, one of two visiting film students – the only non-*campesino* course participants – played the part of 'the soil', while others were representatives of the Monsanto biotechnology company, scientists or local farmers. The group used heroin-pushing as a metaphor to capture the way that hybrid corn was first given for free and only later cost farmers – who had traditionally saved their seed year on year – what they could not afford. Another character, a hybrid corn plant, initially thrives, but later wilts, sending the soil into convulsions and requiring continuous injections of fertilisers from the worried *campesino*. More seed has to be bought from Monsanto, because hybrid corn does not regenerate itself, while new pesticides are needed when plagues of insects invade the new monocrops, sending the *campesino* into abject poverty. Finally, the soil cannot take any more convulsions or injections, and dies. Concluding the play, the soil sits up on the empty stage, and announces "For the sake of *la Madre Tierra*, don't abandon me to this!" It is through such practices that *Madre Tierra* – a figure simultaneously of the traumatised earth and also of a horizon of healing beyond all forms of violence – can both witness more-than-human suffering, and speak from a place where it no longer exists.

What we have seen, then, in this section, is that *Madre Tierra* grounds and anchors a sense of transnational solidarity and resonance with ecological restoration practices – a future orientation that does not fit neatly with what has gone before in the name of 'environmentalism' because it is profoundly at odds with some of the universal ideas at the heart of modern human–environment relationships. The notion of *Madre Tierra* and the emergent

Indigeneity as embedded in Salvadoran permaculture help ground practices of everyday soil repair that are at once linked with agroecological practices in Costa Rica and Japan, and distinct: specifically, they form a basis for coming together to remember histories of violence linked with the Green Revolution, Civil War and colonialism in El Salvador. As such, earth politics in Salvadoran permaculture helps establish a planetary politics that is based in disagreement and critical questioning on the one hand, but new kinds of connectivity and solidarity on the other.

Standing Rock: water is life

Earth politics captures not just disagreements over land-bound elements and ecologies within the Anthropocene, but all elements as they take on new meaning and significance in a radically altered planetary system (see Papadopoulos et al 2022b).[11] In this section, we shift emphasis from soil to water, while maintaining a focus on how this element becomes central to a politics of disagreement (over what worlds are at stake and who can speak for them) and the revival and reclaiming of practices of care and reciprocity. We again foreground an Indigenous politics of water that makes the protection of water inseparable from historical and ongoing settler-colonial violence, Indigenous sovereignty, and the revival and cultivation of different relations to land (Coulthard 2010). Enacting this in practice means learning to be a good relative to water, just as the Salvadoran permaculture practitioners sought to be a good relative to soil. As with soil, Indigenous and non-Indigenous intellectual and activist currents overlap in these new activist cultures, that can be both *in situ* and transnational, traditional and queer, ancient and contemporary.

In April 2016, a small number of women from the Standing Rock Sioux Tribe set up camp in a small valley where the Cannonball River flows into Lake Oahe. They were there to manifest their objection to the proposed Dakota Access Pipeline (DAPL), a 1,172-mile pipeline that was to transport 500,000 barrels of crude oil under Lake Oahe, the only source of drinking water for the Standing Rock Sioux Nation and millions of others downstream (Gilio-Whitaker 2019). The concerns raised by the objectors related to risks of water contamination resulting from a pipeline breakdown or leakage, and because the proposed route of the pipeline was on lands and waters the tribe had never ceded to the US. These lands are culturally and environmentally vital for the Sioux Tribe and as LaDonna Brave Allard of the Standing Rock Sioux Tribe made clear: 'We must remember we are part of a larger story. We are still here. We are still fighting for our lives, 153 years after my great-great grandmother watched as our people were senselessly murdered. We should not have to fight so hard to survive in our own lands' (cited in Whyte 2017: 154).

What became known internationally through a series of hashtags as the #NoDAPL, #Mniwiconi, #Waterislife and #Standwithstandingrock movement quickly became a focus for global environmental and climate activism. The camp at Standing Rock drew hundreds and then thousands of Indigenous and non-Indigenous peoples in solidarity with the Standing Rock Sioux and to protect water and sacred sites. The physical encampment, forcibly shut down by state police and national military forces in early 2017, was linked with solidarity campaigns around the world, led in particular by Indigenous peoples and communities who found their own experience resonate with those of the Sioux Tribe (Monet 2016). The camp opened a space that could hold together, for a time, many otherwise culturally and politically different groups, ensuring that what was at stake was much more than just a pipeline. Though struggles against fossil fuel infrastructures and the logics of extraction they materialise had been taking place for many years prior to 2016, often led by Indigenous peoples, the #NoDAPL movement appeared to mark an important shift in climate and environmental politics.

With a focus on fossil fuel infrastructure, the campaign drew activists and environmental groups whose work centred on climate change (Klein 2016). But, for the Standing Rock Sioux and many Indigenous peoples in North America and beyond, their demands went beyond stopping the expansion of fossil fuel infrastructure. The pipeline was just one more manifestation of the settler-colonial system that had been inflicted on Indigenous peoples for hundreds of years. Thus, resistance from the Standing Rock Sioux was primarily articulated in the name of Indigenous sovereignty and the protection of waters,[12] two struggles that are inseparable from the standpoint of Indigenous political ontology (Woelfle-Erskine 2019).[13] As in the demands for Rights for Nature in Latin America, this is a concern to translate reciprocal relations long understood as sovereignties.

Settler-colonialism is not a one-off event but an ongoing, systemic process (Simpson 2017; Whyte 2017). The spectacular opposition at Standing Rock can even detract from the more insidious, slow and less perceptible forms of settler-colonial violence that leach away the life-giving capacities of waters and lands in Indigenous territories (Nixon 2011) – such as the uranium mines that have been left open in western South Dakota since 1978 (see Voyles 2015). This is just one toxic legacy that is less spectacular than a pipeline, harder to organise around and remediate (Beckett and Keeling 2019). In a context where such histories are actively forgotten and erased, the process of finding ways to express them becomes a vital part of decolonisation. As Dakota scholar, Waziyatawin Angela Wilson, puts it: 'part of the key to our future liberation ironically rests inherently in our stories of suffering' (cited in Brown et al 2007: 20).[14]

In June 2017, just months after the Standing Rock camp was shut down, Chas Jewett, a Lakota–Dakota Indigenous American from South Dakota, and organiser within the #NoDAPL campaign, visited Ireland. She was invited to an event in Cork City on the themes of water and the commons, in a context of recent national protests and mass mobilisation against the introduction of domestic water charges. At the event, she offered a different way of speaking about water and water activism. Drawing attention away from recent European experiences of austerity and sites of (largely) urban activism, she recalled longer histories of colonial violence against Indigenous peoples and deeper considerations of society's relationships with water. She began by speaking about the many forms of systemic violence inflicted on her, her native people, and the lands in which they continued to live – at the scale of the body, the community and the planet, a culture of domination and disregard. Struggles over water were thus placed within a wider and longer narrative of oppression, which is both gendered and racist (Jewett and Garavan 2019).

For Chas Jewett, one of the most positive outcomes of the movement at Standing Rock was the return of Indigenous ceremonies, bringing many native young people back to political engagement (Bresnihan 2019). 'Our bodies hold on to trauma. Our bodies are the repository of all that we haven't let go. So these Ceremonies are a way for us to really let go' she writes (Jewett 2017), emphasising that ceremonies were part of the process of healing and repair, at collective scales. One ceremony that held significance for her was the water walk, or *nibi* walks, based on Ojibwe Ceremonial Water Teachings. In these walks, women carry pails of water taken from the water source and follow the river's course, making offerings of tobacco, praying to the water and singing the water songs. It is the women who do this, because water is a life giver and women, as life givers also, are the protectors of the water (Jewett and Garavan 2019). The walks can be long – Chas Jewett completed one the previous autumn, beginning in Canada and following the Missouri River deep into the US. Through walking, they are there to 'heal the water, to heal themselves, by loving the water and by praying with it' (Jewett and Garavan 2019). As with the *in situ* practices of soil permaculture in El Salvador, the walks and the healing process they enact are about more than repairing waterways. For Waziyatawin Angela Wilson, they are an embodied practice of spatial justice, awakening 'the possibilities of land reclamation for our nation, about the recovery of a way of life that made perfect sense only in our homeland, and about the meaning of justice' (Wilson in Brown et al 2007: 18). As we saw in Chapter 4, such witnessing practices embody a powerful means for speaking in the name of those not present, of bearing witness to past and ongoing injustices.

The dual meaning of healing, as both learning to relate to water with reciprocity and love, and remembering the settler-colonial structures that prevent such relations flourishing, inform what it means to be a good relative to water. When Chas Jewett spoke of water, she spoke of it as a relative, and the 'intense responsibility' that carries with it. In their beautiful, edited collection, *Downstream* (2017), Dorothy Christian and Rita Wong write, '[t]he challenge to reimagine ourselves beyond our skins, as a living part of a larger watershed, can hold both frustration and promise' (Christian and Wong 2017: 7). The ontological standpoint conveyed here – which is similar to that put forward by Rights for Nature proponents in Bolivia and Ecuador – moves us away from the idea of sovereignty as possession, towards the idea of sovereignty as responsibility and care, which involves taking care of lands and the elements within it. Within this perspective, reclaiming sovereignty demands reinvention and rearticulation of Indigenous Land practices and science, as well as ending legal, economic and environmental systems of marginalisation and dispossession (Whyte 2017).

As in Cochabamba and El Salvador, Indigenous/non-Indigenous formations also emerged in Standing Rock, reflecting the fusing of old and new traditions and technologies. Many of the startling images of the camp at Standing Rock that circulated around the world were taken by photographic drone operators. These surveillance technologies were used to bear witness to the scale of state resources amassed against the camp, as well as the magnificence of the natural landscape. They demonstrate the enduring importance of witnessing as a form of activism, as well as the way that novel technologies – even a military technology like drones – can be reappropriated to these ends (see also Millner 2020). The images produced with drones were instrumental in raising awareness and funds for the campaign, reporting human rights abuses and generating a counter-narrative to state and corporate media. Subverting the familiar power relations of surveillance, the use of drones by the water protectors has been described as 'an indigenization of neocolonial military and corporate surveillance technology', as well as a form of 'aesthetic protest in which the beauty of the water, land, and the movement itself are on full display' (Keene and Hitch 2019: n.p.).

As many decolonial scholars have emphasised, we (especially non-Indigenous, Western scholars) should not celebrate and enjoy Indigenous ideas of kinship without foregrounding Indigenous sovereignty.[15] This reiterates a point found in the decolonial movements in the Third World (for whom national sovereignty, supported by regional and transnational organisation, was essential for decolonisation), as well as in the demands of the Zapatistas, MOSOP and LVC. Sovereignty is not just a question of legal control, it extends to include all the material, technological and cultural resources required to collectively adapt and produce viable futures in the face of climate change (Ajl 2021). The return of Indigenous lands will not in itself be enough to ensure Indigenous self-determination – as suggested earlier, these lands need remediation and repair from centuries of degradation and pollution, and communities need access to technologies and resources, and the means to design their own infrastructures for development. This is where

Indigenous science becomes so important, fostering ways to live well with others again in environments that are degraded and toxic. As Max Liboiron puts it in relation to their work on marine plastics in Newfoundland: 'Purity is not an option here – plastics are already in Land relations' (2021: 109). Decolonising science in this context means learning to be a good relative not only to plants and soil, but also to plastics, a question that is directly informed by Indigenous ontology and ethics, while at the same time confronting the toxic legacies of industrial capitalism and settler-colonialism. Liboiron is drawing on a longer tradition of feminist epistemology which moves beyond critique of 'Western' scientific rationality, seeking instead to make explicit and strengthen situated forms of knowledge production, grounding truth in the conditions of everyday life.[16]

As with experimentation in action-oriented pedagogies of agroecology and permaculture, Indigenous science is not just about reviving traditional practices. This is in part because the recuperation of traditional knowledge is not just a cultural reclaiming, it is wedded to material conditions and the practical needs to provide food, nourish and care for bodies and land. Extending this perspective beyond Indigenous contexts, Dimitri Papadopoulos contends that a decolonial politics of matter is 'a politics that challenges epistemic coloniality by transforming materially everyday ontologies of existence' (2018: 172) – including the use, appropriation and hacking of technology, plants, infrastructures, microbes and so on, to affect material changes in everyday existence.[17] Extending decolonial epistemic practice in this way does not equate with the flattening of different contexts, specifically those prevailing in settler-colonial states.

Conclusion

The Anthropocene concept carries in its etymology the universal category of 'man', now elevated to a geological force. But in doing so it has also opened a critical space for challenging this foundational conceit, encouraging different narratives on the origins of current climate and ecological breakdown. Pitched against the technological determinism and simplistic progress story narrated by Globaïa and the UN+20 Summit, decolonial scholars in the present root the current ecological crisis in the conquest of the Americas, the decimation of Indigenous peoples and the development of colonialism (Davis and Todd 2017). Linking the Anthropocene and colonialism means bringing Indigenous (and Black and racially othered) demands, knowledge and voices into the heart of debates over the Anthropocene, as well as linking ecological breakdown to the ecocidal ideologies of colonialism and capitalism. As we have outlined in this book, this is not an entirely new argument. What is significant is that in the second decade of the 21st century, such arguments began surfacing with new force, via movements such as Standing Rock, connecting struggles around settler-colonialism, Indigenous sovereignty and climate change in novel ways. Central to this was the articulation of Indigenous relations to land, an *earth* politics that centres on a disagreement over what is at stake in environmental conflicts.

Yet, the question of how to engage with the specificity of Indigenous demands for sovereignty is not simple. It is important to be clear: the demand to stop fossil fuel infrastructure, or keep fossil fuels in the ground, is not the same as the demand to return settled land to Indigenous peoples (Whyte 2016).[18] In the context of climate justice and urgency around environmental problems, there is a danger that the particular and unsettling demands of decolonial justice could (again) be obscured and incorporated into an undifferentiated environmental or climate movement. Non-metaphorical decolonisation requires the return of sovereignty to Native peoples, including possession of lands that have been taken from them, not mere lip-service to cultural difference. Solidarity of non-Indigenous civil rights and social justice activists with Indigenous peoples is thus no easy task and will remain 'an unsettled matter that neither reconciles present grievances nor forecloses future conflict' (Tuck and Yang 2012: 4).

At the same time, Chas Jewett exclaimed to the assembled audience in Cork, 2017, that "we are all Indigenous". She was responding to a question about the relevance of her account about Standing Rock to the situation in Ireland. Though Ireland has had its own experience of colonialism and dispossession this was not exactly what she was referring to. It is becoming harder and harder to avoid confronting the scale and intensity of ecological damage that extends, albeit unevenly, across the planet.[19] The work of learning, recuperating, reinventing and resisting is not for Indigenous peoples alone (Red Nation 2021). As with the transnational networks that undergird the Salvadoran soil repair practices, so too are the water walks and other ceremonies travelling to other places, weaving with other cultures and struggles in defence of land and water. A decolonial aesthetics does not just mean attending to marginalised or absent voices, but how, when, where and through whom, Indigenous experiences and concerns are made to appear and to what ends. Such an approach keeps open a critical space for fostering alliances across different political strategies and ideas that have been emerging in recent years.

In the examples we have covered here, earth politics form a direct challenge to some of the ways that modern environmentalism had become wedded to managerial and technocratic solutions in two important ways. First, what we characterise as earth politics are social movements and sites of political disagreement that seek to reveal the ways that stages of modern politics – specifically, forms of modern politics seeking universality and claiming to bring together all peoples around common notions – exclude and dispossess by discounting non- and alter-modern forms of knowledge and knowing. The appearance of earth-beings in political claims forms an important way in which this disagreement is made known. Second, and relatedly, earth politics grounds alternative forms of collective politics and translocal association that form the basis for new kinds of collaboration, not based in universality, but pluriversality. It is through such collective associations that we see the emergence of collective figures of politics such as *Madre Tierra* and *Pachamama*, as well as the grounded forms of everyday politics and ecological repair that they make possible.

Fourth Interlude: Making Things Resonate

Surrounding the political interventions that refresh and enliven environmentalism, or push beyond environmentalism into new framings, we notice creative interventions within artistic, activist and scientific fields. In terms of cultivating resonance, as we will argue in Chapter 6, these interventions are able to stir up questions and surprise us, by combining and juxtaposing different sites, historical events, objects and practices. A vital part of resonance is thus the 'shimmer' of difference and distance between bodies – the sense of connection emerging between disparate things and collectives, as opposed to forms of thought that work by implying sameness (for example, recognition). We come to notice how people and places are indebted to one another or are intimately associated in ways we do not necessarily appreciate.

The artists recounted here, and in this book generally, do not simplify these connections, or flatten the differences between geographies and contexts. The Syrian War was not caused by the Svalbard Seed Vault (see Chapter 5); the struggle in County Mayo against Shell is not the same as that experienced in the Niger Delta. But just as they are careful to avoid oversimplification, these works do not fall victim to overcomplexification. Climate change; the Anthropocene; the sixth mass extinction: these are vast processes which, when considered at the planetary scale, tend towards the evasion of responsibility and the exhaustion of meaningful agency. For the artists we collate here, as with many activists involved in everyday organising and making of alternatives, agency lies in experimental practice and the construction of new collectives.

The following examples are not exhaustive (see also Gray and Sheikh [2018] and the associated special issue, as well as Demos [2013] for other important examples) but are the work of several artists that have inspired the work of this book and are mentioned in its chapters. We include references to sources to allow readers to find out more about these artists and their projects in their own terms.

Seeds of change: a floating ballast seed garden (Bristol), 2012–2016 (Maria Theresa Alves)

'Seeds of change' was a migratory project that was created anew in a number of port cities by artist Maria Theresa Alves in collaboration with archaeologists and other local artists, academics and storytellers.[1] The project took the principle of 'ballast', used to weigh down ships in colonial times, as a basis to lead archaeological digs around harbourside areas in port cities heavily involved in the slave trade. The seed garden was a barge later planted with seeds found to have been 'dumped' in the ballast, and thus accidentally spread because of maritime journeys linked with trade of colonial products and journeys including the passage of slaves. The barge was used to host workshops including local schoolchildren – Naomi was involved in co-producing a series of stories with a local storyteller imagining two fictional crossings of the ocean with child protagonists, exploring the spaces of the sea, botanical journeys of the plants on the barge and the different marginal geographies that connect them.

Civic Laboratory for Environmental Action Research

The Civic Laboratory for Environmental Action Research (CLEAR) is a feminist, anti-colonial, marine science laboratory based in Newfoundland, the ancestral homeland of the Mi'kmaq and Beothuk.[2] CLEAR is about doing science differently, not replicating colonial ways of producing knowledge. This means fostering methods that 'foreground values of humility, equity, and good land relations' (CLEAR nd). The extensive archive associated with the project needs to be explored to understand what this kind of careful and creative scientific practice looks and feels like. Humility is key in breaking with the universal assumptions of colonial science: the idea that one set of knowledge practices applies to all places and all times. Place matters; context matters; relations with other human and nonhuman beings matter in all scientific knowledge production. Situating knowledge means taking account of these things that matter, embracing that knowledge can be partial and objective at the same time. This is an ethos adopted from Indigenous and feminist traditions. Begun in 2017, CLEAR resonates with movements discussed in Chapter 5, particularly the resurgence of Indigenous politics (especially in North America) and the important place of experimental practices of science and technology within 'earth politics'.

T.U.R.F. (Transitional Understandings of Rural Futures), 2013 (Deirdre O'Mahony)

On her website, Irish artist Deirdre O'Mahony describes one of her films (*Airtime*, 2012) as 'giving voice to individuals impacted by international environmental policies' (O'Mahony nd).[3] Specifically, she is referring to conflicts over turf-cutting which erupted in Ireland in the late 2000s. Irish and EU biodiversity legislation sought to protect raised bog habitats, in the process preventing the historical practice and customary rights of turf-cutting in parts of rural Ireland. O'Mahony's work challenges the stubbornly persistent, colonial and metropolitan ways of seeing the west of Ireland, particularly the bog landscapes – as environments without people, for aesthetic consumption or resource extraction (including as 'ecological service providers'). In centring the voices of those who are most affected by environmental policies, the turf-cutters, she does not romanticise them, nor seek to find well-formed political expression. As described in Chapter 4, the 'right to subsist' became a key terrain of struggle from the late 1980s. Struggles over land and territorial resources were linked to the expansion of traditional commodity markets (for example, NAFTA Agreement), as well as the consolidation of global environmental management – the protection of 'environmental resources' through the exclusion of Indigenous and rural communities.

CLIMAVORE

CLIMAVORE is a long-term collaborative project that sets out to explore aspects of food production and consumption in relation to landscapes transforming through climate change.[4] CLIMAVORE is a creative concept posed to constellate these interventions, thought of as a form of devouring that ties food production to forms of pollution and devastation. Framing our diet within a globally financialised landscape, and challenging large-scale agribusiness groups dictating what is to be produced and consumed, the notion of CLIMAVORE critically questions the geopolitical implications behind the making of climate alterations and the pressures they enforce on humans and nonhumans alike. On the other hand, the artistic interventions are careful to attend to the ways that nonhuman animals, material landscapes and human–nonhuman collectives are not only changed by these relations, but also active agents in the recomposition of alternatives, and participatory dimensions help illuminate alternative future trajectories. The gallery on CLIMAVORE's website reveals the ways these concepts have been explored in landscapes from Skye (Scotland), to Istanbul (Turkey) and Los Angeles (US). In each location CLIMAVORE collaborates with experts in ecology, marine biology, agronomy, nutrition and engineering, among others, to create dynamic, site-specific performance-sculptures that pose fresh questions as part of their interventions. For example, in Skye, 'On tidal zones' was an installation based on an underwater oyster table that formed the centre of discussions around intensive salmon aquaculture and climate change, around meals served up by local chefs participating in the discussion.

6

Conclusion: Resonance Beyond Environmentalism

> If toxic air is a monument to slavery, how do we take it down?
> Forensic Architecture, in partnership with
> RISE St James (2021: np)

In early June 2020, the statue of Edward Colston, a 17th-century slave-trader (1636–1721), was pulled from its plinth and thrown into the Bristol docks (Antink 2020). The historic moment was captured on many phones and shared widely on social media. Cries of jubilation, people jumping up and down, screams between anger and delight, capture an atmosphere of catharsis: the release of emotions otherwise hidden from public expression. There had been long-running campaigns to remove the statue, as well as Colston's name, from streets and other public spaces. Bristol, a port city, was established through the significant profits derived from the slave trade, and Colston was a key figure in that economy. What happened in the summer of 2020 was a spectacular reckoning with colonial, racist history, sparked in turn by the murder of George Floyd in Minneapolis, thousands of miles away. Before sending Colston off into the harbour, protestors knelt on the neck of the bronze statue, invoking the death of Floyd whose last words, "I can't breathe", had become a powerful slogan for the Black Lives Matter (BLM) movement around the world.

As BLM protests, which began before the Colston statue was pulled down, spread beyond the US, the dismantling of public memorials to histories of slavery and white empire became its own trend. People around the world connected racist policing in the current moment with longer legacies of colonial violence still memorialised in cities and public spaces.

That summer also saw historic wildfires ravage across North America. Whole towns were engulfed in flames, and apocalyptic images of climate breakdown filled the media. It shouldn't be hard to connect the sudden manifestations of long brewing climate and ecological breakdown with the colonial histories being protested loudly elsewhere – but unfortunately *these* connections are rarely drawn. Indeed, while each new and intensifying wave of ecological crisis – whether COVID-19, unbearable temperatures or megafloods – is greeted with horror, it tends to be met with a doubling down on technical and economic innovations that ensure nothing really has to change. With growing urgency around climate and biodiversity breakdown, it seems there is less space, but more need, for an environmental politics that reconnects critique with vibrant anti-racist and anti-colonial movements, in line with historical examples we have traced in this book.

Fortunately, there are new efforts and movements engaging in precisely this kind of work.[1] Such collectives enact aesthetic labour, working to show up the excluded connections and the resonance that exists between diverse movements and concerns. For example, Forensic Architecture (FA),[2] a research agency based in Goldsmiths University of London, recently partnered with fenceline community activist group RISE St James, to resist the proliferation of petrochemical industries in St. James Parish, Louisiana. RISE St James had objected to plans to build a massive plastic factory on the banks of the Mississippi River in St. James – an area known as 'Cancer Alley' or 'Death Alley' because of the many petrochemical refineries and associated industries already based there. As a primarily Black community already unjustly burdened by industrial polluters, RISE St James follows in the footsteps of decades-long struggles for environmental justice in the US. But the history of environmental racism stretches back much further than the 1970s, to the cotton plantations and the violent systems of slave labour. Such histories are highlighted in the short film that accompanies FA's investigations with RISE St James, which opens with the removal of a bronze statue of a slave trader (Forensic Architecture, in partnership with RISE St James 2021).[3] Alluding to George Floyd's powerful last words, the narrator asks: "If toxic air is a monument to slavery, how do we take it down?" The film, which is only one part of the collaboration, makes clear often hidden connections between environmental devastation, race and colonial capitalism. This also culminates in clear demands that articulate between past, present and future: an end to the expansion of the Petrochemical Corridor; provision of labour opportunities that sustain communities in place; and ecological reparations, for damage inflicted over centuries on air, soil and water.

The methods FA deploy are classified in their own terms as *forensic*: the presentation of systematic evidence to address wrongs. Forensic inquiries reveal the hidden and seek modalities to make that appear to audiences with whom it might not currently resonate. For the film to have effect, it must reach new audiences and draw connections with linked struggles. Meanwhile, the investigative work to surface relations between pollution, social and environmental justice, and the structural legacies of colonialism opens up space for connections to be drawn between the localised struggle of RISE St James and other sites and subjects. In turn struggles elsewhere find new momentum and aesthetic frames and figures to make their claims.

We refer to the drawing of such aesthetic connections as *resonance*. Resonance describes how social and environmental movements continually expand the terrain of what is at stake in environmentalism. That is, while fighting for objectives (the defence of land, a change in the law, public awareness of chemical pollution or the effects of toxic waste on bodies), they establish connections with linked movements and struggles, making clear that one cannot be achieved without the other. Practically, this involves drawing new links between ecological damage (or reparations) and the diverse ways that structural injustices play out in the lives of individuals and populations. Yet resonance also has what we called, in Chapter 5, pluriversal qualities. Rather than referring us back to a single issue or a unified position, resonance makes room for distinct claims and demands. Resonance within environmental politics thus strengthens diverse forms of struggle and inquiry, rather than the building of momentum around a single way of framing and solving a problem.

In this book, we have touched on multiple instances of resonance thus described, historical moments where traffic between intellectual and activist communities was lively and experimental. Part of the motivation for the book has been to illuminate this excess, the kind of ferment and energy that subtends a broader set of movements, making their claims widen and deepen. This background to movements and ideas is often not remembered, or rather, is erased violently from the historiography of modern environmentalism, which has tended to prioritise heroic narratives and single-issue struggles. This, in turn, has served to make environmental struggles seem separate from movements and claims that co-emerged with them: anti-racist, anti-colonial and feminist, for example. In drawing this book to a close we want to make clear the cost of divorcing environmentalism from this wider field of resonance – and the opportunities for reconnecting with it.

In this chapter, we therefore unpack the concept of resonance, showing how this helps connect the different examples we have looked at in the book into an agenda for moving beyond environmentalism's blindspots. Rather than assuming a universal basis for collaboration, resonance assumes difference as a basis for collaboration. Indeed, resonance *requires difference to operate*; as in the visual and acoustic definitions of the term, resonance works by building a field of connections between elements that are not the same. In this sense, resonance also helps describe the way movements connect and expand one another's capacity, even when this is not formalised or made explicit. Through resonance we can talk about the ferment of the 1960s, when currents of anti-colonial, feminist, worker and environmental movements strengthened one another's claims without being formally connected, as well as about the ways that migrant-worker campaigns against pesticides contributed to the ban on DDT, even though they are not often remembered.

The remainder of this chapter, which also functions to look back on the book as a whole, unfolds as follows. First, we revisit key arguments we have made in this book, following how movements and struggles negotiated the blindspots we have identified in modern environmentalism as it has evolved since the 1960s. Second, we give further definition to our concept of resonance, looking back on the chapters and examples in the book to make clear its importance. This brings us to the heart of the book and its political project in relation to environmentalism.

Movements beyond environmentalism

Today, emphasis in environmental negotiations falls on working towards shared targets, measured by meeting standardised points along fixed scales – something that seems urgent in contexts where climate change and the biodiversity crisis have been met with limited effective intervention by national governments and transnational organisations. Yet some of the dominant ways of making the problem visible and solvable keep us from addressing it. This has been a key argument of this book.

We are making a provocation here: clearly, many environmental campaigners and organisations have worked painstakingly, and often in the face of scepticism and violence, to ensure that ecological issues are taken seriously. These efforts have also brought about important changes within policy making and environmental regulation. We do not wish to underplay this work. What we are arguing, however, is that there remains an aesthetic regime tied up with the name environmentalism that prevents the broadening of struggles beyond a relatively narrow set of concerns. In practice, keeping the focus on this narrow field prevents us addressing the wider causes and drivers of environmental destruction. It also keeps us from seeing the ways that the 'solutions' proposed often reproduce (racialised and gendered) colonial inequalities and legacies.

Here we want to make three points about the character and evolution of this aesthetic regime.

The first is that, from the 1960s until the 2000s, *there has been a shift from environmental movements towards environmental management*. In the 1960s, particularly in North America and Europe, there were new and growing concerns with the environmental effects of industrial capitalism and post-war technologies. In Chapters 1 and 2 we described how the very foundations of post-war economic development and progress were challenged. At a time when expert and popular understandings about the environment were in their relative infancy, there was a brief period when environmentalism was part of the ferment of radical social movements and politicisation in the 1960s and early 1970s. In Chapter 2, we showed how the organising of Filipina and Mexican workers against California fruit growers resonated with the cultural impact of Carson's *Silent Spring*, itself the product of many years of research and advocacy against DDT and other chemicals. While there was not direct collaboration between these movements, the efficacy of the consumer boycott of California grapes in the late 1960s, forcing major fruit growers in the region to enter into a collective bargaining agreement with thousands of workers, was successful partly because of this broader momentum and social context. Ultimately, however, the unevenly distributed risks associated with chemical pesticides were regulated through new forms of scientific expertise and state oversight, suppressing the more radical demands and collective energies of agricultural workers and the public.

While we do not, in this book, cover all aspects of the history of early environmental or allied movements, we note that by the 1990s, when the concepts of sustainable development and environmental governance we unpack in Chapter 4 were becoming consolidated, many of the more radical critiques of industrial capitalism made during the 1960s had been pushed to the margins. As environmental concerns become recognised and increasingly mainstreamed within global governance regimes and corporate strategies, we saw new sites and subjects of disagreement surface. In Chapter 4, we followed the emergence of LVC, the Zapatista movement, and the struggles spearheaded by Saro-Wiwa as examples of the perpetual challenge to the enclosure of environmentalism that continued to take place. Such movements have continued to expand and widen what is understood to be at stake in the expansion and development of capitalist systems of production and consumption. They also innovate conceptually, rejecting the idea, for example, of 'environmental resources', and devising figures like *Madre Tierra* to articulate the relationality of ecological systems with livelihoods, cosmologies and human history. In like manner, the permaculture practitioners in Chapter 5, in dialogue with agroecologists, conspired to bring labour and livelihood to the heart of their ecological practices, while the collective of Indigenous and allied activists at Standing Rock combined technical tactics with ritual in a more spectacular form of territorial defence. In that chapter we emphasise how the modern binaries (of nature versus culture, of human versus nonhuman, of modern versus tradition) underpinning modern environmentalism are radically questioned within such allied movements, as well as the notions of universal human community that ground global environmental summits and agendas. In place of consensual backdrop emerge concepts of pluriversalism, and the idea of 'earth politics', which wants the earth, and the possibility of coexistence of its multitudinous species and peoples, albeit on very different terms.

A second, related point is that *the shift from environmental movements to environmental management consolidated aesthetic tendencies already present in*

Western modes of development and culture. As we noted in Chapter 2, the settled, agrarian landscapes that recur in *Silent Spring* and idealised depictions of California reflect a white-settled vision of America that excludes the exploitation of migrant labour, Indigenous dispossession, and the many struggles undertaken to resist these ongoing injustices. The spatial imaginary of the environment inherited from settler-colonial ideology was always limited and posed problems for raising multidimensional and intersectional critiques concerning race, colonialism and capitalism. As we have argued throughout the book, modern environmentalism leans heavily on spatial imaginaries of wilderness conservation/preservation and empty land [*terra nullius*] closely allied to white nationalism and colonial imaginaries. These spatial imaginaries are central to the legitimisation of other processes quite distinct from conservation, including the making of polluted atmospheres, the dumping of toxic waste close to Black, poor and/ or Indigenous communities, and the use of particular sites for the testing of military weapons (Liboiron 2013). Drawing on high modern principles of developmental thinking and intervention to solve pressing issues has not, we argue, made a definitive break with these histories or imaginaries. Indeed, in many ways it has entrenched them into the backdrop of the models that are now understood to be technical and therefore neutral. Thus, when we imagine 'environments', we might imagine retrofitted industrial infrastructures for preserving seed biodiversity (see Chapter 4), global carbon markets and carbon accounting, or serried ranks of offshore wind turbines. We are not questioning the potential value of some aspects of these technologies. What we are emphasising is how these interventions function to render environments abstract and removed from contested social relations and claims, as well as violent histories of exploitation and dispossession.

Yet herein lies our third, and most important, point, which we frame as a question: *why is it that so many movements concerned with ecology and land do not use the term 'environmentalism' or even the 'environment' to describe their activity and claims?* Rabindranath Guha and Juan Martinez-Alier (2013) have addressed this question by contrasting the environmentalism of the Global North, focused on technological innovation and the 'cult of wilderness', with the 'environmentalism of the poor' devised in the Global South primarily out of necessity. When connected to struggles for livelihood and survival in the face of state and corporate incursions, they argue, grassroots campaigns seeking to protect specific environments *do not make demands in the name of the environment*. This helps point to a plethora of 'environment-linked' movements that don't find affinity with the mainstream project of environmentalism. Instead, social movements which *do* link environmental or ecological issues with land, labour, livelihoods and/or histories of colonialism frequently use other names: agroecology, decolonisation, anti-imperialism, self-determination, food sovereignty, earth politics, *Madre Tierra*, territorial defence, permaculture, ecofeminism – rather than environmentalism.

Here it is important to clarify that when we speak of the history of modern environmentalism it is not so much the *language* of the environment that matters. This is important, as some efforts to displace 'environmental' thinking – for example, through an embrace of earth systems and climate governance – can be apolitical, even anti-political. What gets lost in such moves are decades of organising for environmental justice and discussions over how best to organise for reparative futures. Meanwhile, many political groups and movements *do* continue to organise under the banner of environmentalism or environments. Rather our target here is the *aesthetics* of environmentalism: what is illuminated in the name of the environment and how; what is excluded and rendered part of the background. For example, we have followed how particular ways of framing environmental damage result in technical-managerial solutions that lean on the imaginary of one unified world and planet. We have also emphasised counter-examples that have kept questions of labour, livelihood and colonial histories connected with accounts of 'the environment'. As we have argued with the concept of provincialising environmentalism (see Introduction) we don't need to abandon the reach of the 'environment' to create space for alternative forms of environmental politics. What we do need to unsettle are the universal claims embedded in modern environmentalism. The search for pluriversal, rather than universal, modes of operating requires new and refreshed terms, and different ways of seeing.

All three of these points relate to a second key term we have deployed through this book: disagreement. For Jacques Rancière (1999; 2013), a political theorist central to our theorisation of political aesthetics in this book, disagreement marks the appearance of a collective which isn't counted as a 'part' of the social order and the assertion of a polemically different account of things through the presentation of a fresh set of demands ('what counts as environmentalism for you isn't the same as what counts as environmentalism for us'). The disagreement involved is hence the production of a dispute over the 'givens' of a governed order – what and who may appear as the bearer of rights; who or what counts as a political actor; what is considered the 'problem' to be solved. As we have suggested, this politics takes place on the level of aesthetics, which is the domain of seeing and sensing where what can be perceived (or heard) is invested with political significance – or, where 'givenness' becomes framed according to a historically particular 'distribution of the sensible' (Rancière 2013: 13). In terms of modern environmentalism, disagreement concerns the who, how and what of its operations: it engages questions of who or what is included in environmental politics; how (far) a problem is understood as 'environmental', and what is at stake in environmental politics.

Disagreement was at the heart of the global revolt of 1968 that we covered in Chapter 3, where diverse movements presented the Third World as a new world-historical actor, and sought to explore work beyond the factory and wage-relation. While not coordinated around a single, shared political objective, nor directly foregrounding environmental concerns, this proliferation of sites of struggle supported new ways of thinking about, and acting on, ecological crises. Over time, these sites are not always remembered together in a history of modern environmentalism, and colonial and racist biases are partly to blame for this selective memory. Yet, we claim, they were essential to one another, and to building a wider stage for the political claims that were successfully enacted. Thus, it is our choice to remember together movements that were only partially connected in practice.

In Chapter 4, disagreement centred on new discourses of 'sustainable development' and 'globalisation (after the end of history)', with territorial movements like the Zapatistas and MOSOP rejecting these new consensual, global arrangements. The Zapatistas, LVC and Saro-Wiwa were not acting as a unified movement – there is not even clear evidence that they recognised one another's struggles during the 1990s. Nevertheless, the figures each movement established to communicate their struggle, and witness to violence, resonate with one another, and function partly by establishing a wider aesthetic field that is alternative to dominant ways of thinking about land, globalisation and self-determination. Consider the LVC slogan: 'Food Sovereignty is Land, Water, Seeds, Bread and Solidarity!', alongside the Ogoni Bill of Rights demanding a greater degree of political and territorial autonomy, and the San Andreas Accords agreed between the Zapatistas and the Mexican government in 1996 granting autonomy, recognition and rights to the Indigenous population of Mexico.

In Chapter 5, meanwhile, disagreement is played out via the staging of Indigenous politics and the presencing of worlds excluded by, and unrecognised within, extractive economies. The concept of 'earth-beings' is precisely a concept of disagreement, and a break away from forms of political jurisdiction that do not consider the persons of Indigenous cosmologies as such, but rather as cultural or 'primitive' belief. The multiplication of struggles creates resonance for such articulation from one context to another, while figures like *Madre Tierra* expand the fields for collective disagreement and struggle. Furthermore, this move to articulate a pluriversal environmental politics around land and water is not just restricted to Indigenous contexts and politics. This is a consistent feature of decolonial environmental politics – expanding the scope of the problem by building resonance and asserting disagreement, and thereby expanding who and what counts as part of environmental politics.

In the next section we unpack the concept of 'resonance', which articulates how movements may build the field for one another to operate and expand influence, even when they are not directly allied or associated. We have sought to notice how often overlaps between movements exist, even when they are not commonly remembered together. We are not suggesting that these movements are allied with one another, or even that they should have been. Resonance is a term to articulate the way these groups widen fields of agential possibility without falling back on the idea of a single shared world and a single modality of saving it. Rather than amplify consensus, resonance builds concern in connection with questions and objections. In this sense, it has, as a form of aesthetic praxis, been crucial to alternative environmental politics as a form of disagreement.

Resonance beyond environmentalism

In the context of this book, resonance, or 'making things resonate', describes how the social and political movements we have followed amplify concern among linked or sidelong movements, even when they do not initially share a common demand or mode of organising.[4] In relation to the making of modern environmentalism, we suggest it offers a valuable way to move beyond blockages that we currently face. To make this clear, we will now develop what we mean by resonance, drawing out several statements from these ideas, and revisiting examples of resonance from the book so far.

Resonance is a musical term that invokes the idea of making bodies in close frequency vibrate together. Musically, it relates with the word 'to resound' and can be understood as the source of sound production in most musical instruments. Musical instruments are constructed to make sound waves bounce off surfaces and air columns to achieve particular frequencies. In this sense, resonance means *the oscillation of a system to create a greater amplitude*.

In the context of the book, resonance refers to the way events, figures, ideas and depictions of struggles can resound between different places and contexts in a way that amplifies their concerns and demands without making them the same. As we have shown, movements may resonate with each other, strengthening one another's momentum, even where they are not directly linked or in collaboration. It is important to note, however, that *not resonating* does not mean failure or vice versa – there are many examples of important artistic works that found their audience many decades later, or of historical figures and campaigns that took on more force and achieved greater gains in their future remembering. Resonance sometimes works *across* time periods (diachronically) rather than between places in one moment (synchronically). Understanding how resonance works creates an alternative way into 'environmental' issues, revealing important clues to the way that ecological concerns can – and need to – be articulated with other claims to avoid becoming captured into 'ways of seeing' that reinforce one, overarching framework. Moreover, resonance also encourages us to appreciate that the language of *listening* is vital for participating in a politics that requires perpetual reinitiation, humility and openness, rather than over-confidence in what should be done.

Importantly, resonance is not to be understood as a normative concept; it does not necessarily result in more progressive projects and politics. Resonance can just as easily result in new right-wing formations and alliances – for example, between disaffected working-class communities and elite corporate interests, as has happened on both sides of the Atlantic in recent years. But resonance is necessary to foster more just eco-social futures because it fosters polyphony, connection between incommensurable positions. Where this polyphony exists and is fostered, and where the conditions are fostered for acts of reclaiming in the name of justice and wrongs to be articulated, there is potential for more equitable and reparative futures. Resonance points to a capacity to exceed simplifications and sameness, and a potential to interrupt dynamics and universalist claims. And this, at least, is an important orientation for any progressive politics aspiring towards a better world for the majority.

Resonance is, then, a *means of amplifying a matter of concern, by increasing the capacity of bodies to care and act*. It is not possible to build resonance without a field of difference, and in this sense, resonance has a vital relationship with incommensurability, as developed in the previous chapter. Bodies that already are completely in tune with one another cannot 'resonate'. There must be a gap in between them. Jennifer Wenzel (2019) describes this as the gap between 'here and there' that remains ambivalent due in large part to the 'uneven vulnerability' (Nixon 2011) of planetary ecological breakdown. This uneven vulnerability is often met with 'readymade responses of uncritical paternalist sympathy or a too-easy sense of solidarity or shared vulnerability' (Wenzel 2019: 9). But it can also be traversed by forms of 'world-imagining from below' where marginalised subjects are able to 'situate their precarious local condition within a transnational context' (Wenzel 2019: 9).

This gap, and efforts to traverse it from below, are essential to the modality of the pluriversal systems we wrote of in Chapter 4, which, in contrast to forms of organising that seek to establish universal commitments, explicitly value disagreement and incommensurability. Thus, resonance is a helpful term for acknowledging the field produced by different movements operating in the period of 1968–1974 we covered in Chapter 3, when environmental, worker, feminist, anti-colonial and Third World movements were acting on similar terrain but occupying very different positions on that terrain. Similarly, the way public concerns in the 1960s over risky chemicals helped to create a context in which migrant agricultural labourers and middle-class consumers could connect even as their social position, experiences and stakes were far removed from one another. This distance caused problems when the demands of workers, consumers and environmentalists in the industrialised North were separated from the demands of social and political movements across the Third World. The same when the concerns of privileged Californian consumers were addressed in ways that did not address the continued exploitation of fruit pickers in the fields. We can use the word resonance to describe how these movements amplified one another's concerns, never forgetting that different concerns are always inscribed with privilege and power.

The vibratory and acoustic qualities of resonance as a term further encourage us to think about the ways that this can sometimes be felt in the body, detected or picked up like music, before it can be discerned into language, words or claims. We like the way this allows for the way that different movements may have been emerging at the same time with complementary rhythms or melodies, albeit with quite different content, claims or backgrounds. In keeping with this analysis, the previous chapter followed practices of experimentation that render everyday ecological and activist practices livelier, opening potential for more-than-human collaboration and solidarities between diverse collectives. The interludes have shown witnessing practices and resonance that constantly exceeded solidifying practices of environmental management. And, at points throughout the book, using the term resonance we have noted the way that movements have acknowledged, recognised, showed gratitude, reactivated and/or caught the flow of earlier events and movements. It has been important to us to emphasise that environmentalism has not had one linear history – at their richest, and most dynamic, environmental movements have rejected simple progress narratives and fix-it solutions and have embraced multiple temporalities.

It is useful to contrast resonance with the now popular idea of the echo chamber. The echo chamber is an enclosed space where sound repeats and bounces back *without difference*: all that comes back is the same. In terms of making political claims and organising for environmental protection, an echo chamber is a *modus operandi* that takes place by repeating and strengthening common claims, rather than making room for what we have called incommensurability. Incommensurability articulates the impossibility of aligning diverse ways of knowing environments and our relationships to them; it is a principle that makes difference primary and essential to collective acts for caring for ecologies and worlds. In keeping out incommensurability, or failing to acknowledge it, the echo chamber keeps things repeating according to predefined patterns, even as it may appear to offer new solutions.

Today, the echo chamber is often used to describe the digital monocultures cultivated on corporate owned social media platforms. Despite an almost relentless frequency of social interaction, and even the feeling of being exposed to diverse opinions, the algorithmic systems underlying these enclosed platforms ensure we only hear a minority view. Thus, new information and communication technologies function to streamline social groups, keeping them from interacting – although this is not inevitable – leading to forms of repetition without difference. William Connolly (2007) uses 'Evangelical-Capitalist Resonance Machine' to describe a similar phenomenon, where neoliberal capitalism and mediatised religion co-articulate with each other in closed circuits. Here, different concerns are captured and made to reverberate with one another like electrical charges, 'engendering a machine larger than the separate interests, narrowly described, of the parties' (Connolly 2007: np). With other movements and literary traditions, we used the term 'witnessing' in the third interlude to describe the labour of presencing in connections and effects excluded by the 'global resonance machine' of sustainable development and environmental protection. Critically, such practices operate by establishing resonance for concerns that are not currently audible beyond a narrow setting. While the history of modern environmentalism in many ways consolidates an echo chamber that had been established through Western modes of development and 'civilisation', there are numerous examples of resonance breaking in. Sometimes these acts took place in the name of the environment, and sometimes they did not.

The Zapatistas were quick to identify the power of state and corporate-owned media for the embedding of neoliberalism as part of the making of an echo chamber. But they also sought to interrupt that by creating their own narratives, media networks and practices of resonance. They recognised the degree to which privatised 'public' media shaped public opinion and justified state violence. The media narratives lauding free trade, consumption-led progress and development had allowed President Carlos Salinas de Gortari to claim that Mexico was entering the 'First World' with the signing of the NAFTA Agreement in 1994. Erased from this account were the Indigenous peoples in the Chiapas who were (once again) hurled into the dustbin of history. The majority only appeared in mainstream media when they were killing or being killed – the 'part without a part', as Rancière (1999) would call it; those who have no say in political life because they are judged to be without the capacity for intelligent speech. The Zapatista response to their exclusion from mainstream media was the creation of their own stories and narrative forms which, supported by independent media collectives and infrastructures, including the still relatively open internet, became a key strategy for witnessing from below throughout the 1990s. Like the collectives in Chapter 5, they also link this practice with a pluriversal ethos: a concern to establish the means for multiple worlds to thrive.

Conclusion

Monuments to slavery are afterlives of colonialism. They remind us of all that must be forgotten for them to remain standing. FA and RISE St James draw our attention to other monuments to slavery and colonialism, such as toxic air. We can add rising sea levels, species extinction and deforestation. To topple these monuments requires moving beyond the partial and privileged ways of seeing that have persisted within modern environmentalism.

While the environmentalist way of seeing is not *the* logic that enabled certain bodies and places to be designated as sites of experiment for Green Revolution technologies or toxic dumping, as we discussed in Chapter 1, we have worked to show how colonial geographies and capitalist modes of accumulation have been further enabled through the consolidation of modern environmentalism. This continuity, as we have framed it in the book, is made possible through, first, the insistence on preserving and imagining landscapes without people. Second, through the consistent erasure of histories of struggle over livelihoods, land and self-determination, allowing incoming plans and models to be imposed without political debate. Third, through the uncritical relation to histories of colonialism, leading to forms of intervention that are top-down and paternalistic, or which legitimise particular sites and populations as acceptable casualties of progress. While many of these assumptions *have* been critiqued avidly, leading to participatory models of conservation and environmental protection, for example, environmentalist ways of seeing have continued to provide aesthetic consistency to neocolonial forms of solving ecological problems. This means that the 'solutions' themselves perpetuate knowledge and power hierarchies established through colonial violence, even while seeming to solve the (ecological) excesses of industrial production.

In this book we have followed the emergence of movements within and beyond the framework of modern environmentalism, that have demonstrated how to presence back in excluded issues of race, colonialism, patriarchy, labour and more. Often, we have found, such movements and collectives prefer to organise around other terms to achieve this: food sovereignty, agroecology, earth politics, Indigenous sovereignty, land justice, ecofeminism, national liberation, self-determination. However, the concern to protect and restore ecological systems and species diversity is not displaced by these concerns. We do not wish to make an argument that would suggest that all that is 'environmental' should be deprioritised behind other socio-political concerns. The argument is the opposite: modern environmental politics is and has been at its richest when it resonates with these wider movements. To see environments *without* attention to questions of justice and injustice, histories of colonialism, and dynamics of capitalism is to perpetuate such systems. The 'movements beyond environmentalism' in the book's title are not movements away from ecological concerns, therefore, but movements that connect with other movements operating in a field of resonance, embracing difference where necessary, establishing disagreement where it is called for. These movements and this momentum have been present since the birth of environmentalism, and can be witnessed today.

Coda: Afterlives

We have also heard and seen the resistances and rebellions that, even when silenced or forgotten, do not cease to be vital indicators of a humanity that refuses to follow the system's hurried pace toward collapse. ... They show us that the cure, if there is one, is global; it is the color of the earth, the color of the work that lives and dies in the streets and barrios, oceans and skies, hills and valleys—like the originary maize, it has many colors, hues, and sounds.

>Part Six: A Mountain on the High Seas
>(communiqué from the Indigenous Revolutionary
>Clandestine Committee General Command of the
>Zapatista Army for National Liberation, October 2020)

These words were written by Subcomandante Moisés, General Command of the Zapatista Army for National Liberation, in October 2020. The poetic style and generous embrace of difference-in-struggle should be familiar to anyone who has read Zapatista communiques over the past 30 years. The context of writing is very different, though, to the one we explored in Chapter 4. The COVID-19 pandemic; the increasing regularity and intensity of climatic events; the imminent possibility of biodiversity collapse are different symptoms of systemic breakdown on a planetary scale. As one response to this emerging scenario, the Zapatistas announced they would leave the Jungle and come to the world. This was the basis of their 'Journey for Life' – an expedition of Zapatista representatives across the Atlantic, and ultimately to five continents. Reversing the original expedition of European conquest, the 'Journey for Life' was not about obtaining compensation or seeking apologies from Spain. Rather, the Zapatistas made clear that they were journeying to meet others in resistance; others who were also fighting for life against the patriarchal, oppressive and exploitative system of global capitalism.

As part of the 'Journey for Life', a contingent of Zapatista men and women arrived in Ireland in October 2021. In advance of their arrival there was a flurry of organising by individuals, activists and community groups keen to meet with and host them. Involving people across the island of Ireland, of different ages (some learning of the Zapatistas for the first time, others remembering how they had been inspired in the 1990s), and engaged in different forms of collective politics, the ten-day 'Zapatour' created an energy within social movements that had been missing in Ireland for some years.

One exchange in Ireland took place between the Zapatistas and a community campaign in the Sperrins, a rural region of County Tyrone, close to the border. The community is resisting a proposed gold mining project led by Dalradian, a Canadian mining company. On a rainy, autumnal afternoon, eight Zaptatista women shared the story of the Zaptatistas in their own Indigenous language, translated into Spanish and then English. The Zapatistas also heard about the story of the Sperrins, and how this community, which extended to the mountains, bogs and streams, were asserting their right to live in a healthy climate, to clean water, to clean air, to unpolluted land and sustainable farming.

Six months after the Zapatista visit, in July 2022, 21 Indigenous water protectors from Standing Rock were due to come to Ireland in solidarity with Irish communities defending air, land and water but were prevented due to risks around COVID-19. Described as the 'Making Relatives' tour, the events still went ahead, with communities from both sides of the Irish border organising public meetings, protest walks and social gatherings. These communities are on the front line of mining and other extractivist industries, learning and teaching about their respective histories and struggles. The tour included a '*nibi*' or water walk along the River Foyle, and culminated in a Climate Camp in Kerry, where a 14-year campaign against a proposed liquefied natural gas terminal has been underway. The Climate Camp was the first to happen in Ireland since 2010, and it brought together young and old activists from across the country for five days of workshops, movement building, food, dancing and direct action.

Both the Zapatista visit and the 'Making Relatives' tour brought people together who hadn't been actively in touch for some years, as former campaigns had petered out, and social spaces had closed. The visit connected urban and rural activists and community groups, challenging us to think about why urban and rural concerns are so often disconnected in environmental politics. These encounters helped build new connections and language in the fight against mining and extractive industries in Ireland, including the rights of nature, degrowth and regenerative justice. In a context where these industries are extending their operations globally, driven in part by net-zero policies and 'green' commitments, translocal solidarities within and beyond the island become more and more significant. In the process, these encounters are also helping to resurface fundamental questions relating to the legacies of colonialism, reparations and decolonisation, as they manifest differently in specific contexts.

We close our book with this recent set of events, minor as they may seem, because they capture something of the hope and energy found when past struggles resurface through new encounters and contexts, establishing new fields of resonance. In her book about 1968, the cultural theorist Kristin Ross (2008) uses the term 'afterlives' to refer to the way that historic events, such as those of 1968, reappear in new ways long after the moment is over. In this way, they live on through new kinds of protest and new social movement configurations, reopening currents of disagreement in new ways. In other words, Ross frames the afterlife as an event imbued with potential that can be released in unexpected and vastly different contexts to the one it originated from. When this happens, the event recomposes the fabric of the social into something altogether new. Writing about the 1995 general strikes in France, and the alter-globalisation summit in Seattle in 1999, she writes: 'these events created and continue to create a new optic on 1968. ... It brought an end, that is, to the end of May, by giving it a new afterlife, the contours and rhythms of which are still before us' (Ross 2008: 214–215).

The idea of afterlives stands in contrast with the less hopeful but more widespread idea that events that once held potential for dissensus and disagreement have been incorporated by the state or co-opted by neoliberal capitalism. Afterlives leave open the possibility for the reconfiguration of a dynamic energy that was present in earlier movements, as well as the interaction between movements that may appear at first to have little in common. In the terms of the previous chapter, we might think of this as a form of diachronic resonance, reopening legacies of past eventfulness in the present.

Without this orientation to excess and eventfulness, this book could quite easily have been a narrative without hope: tracing how environmentalism, initially part of a radical historical conjuncture, became progressively trapped in managerial institutions and corporate business strategies inadequate to the moment of crisis we find ourselves in. Our orientation to excess and eventfulness opposes this linear and claustrophobic account, asking for perpetual sensitivity and attunement to the potential for new sites and subjects of disagreement, and new affinities. The movements and figures that we have traced in this book are not always connected or explicitly part of environmental politics, but they are part of a wider fabric of ideas that demands the protection of ecological life as part of a web of land, labour and social relations that must be reconstructed beyond the devastation of colonialism and capitalism.

The movements and figures covered in this book are rooted in specific places, and yet they were also part of a global circulation and encounter of ideas and knowledge that was often surprising – a globalism from below. Where we find inspiration and what we would like the book to contribute to is in revisiting the radical demands and openness of past movements in ways that resonate with concerns and movements today. There is a generosity inherent to these encounters; a resistance to the idea that words and actions have a predetermined destination or role. Such generosity, particularly at times when the path of travel is unclear, does not equate to a flat openness to everything. Nor does the circulation of ideas and practices happen spontaneously. We must work at it.

Notes

Chapter 1

[1] On this point, 'environmental' warfare in the Second World War was the deliberate attempt to contaminate the living systems of an entire zone, nothing to do with protecting living systems (Westing 1985). Throughout the Cold War period there remained a hazy distinction between military and environmental expertise: in the UK, the same Oxford botanists who developed understandings of ecological precarity were those consulting on agricultural defence and informing British military commanders in Malaya (Hamblin 2013). We come back to this critical connection in Chapter 2.

[2] By 'aesthetic regime' we mean the way that the felt fabric of the world is established through repetition and sensory training. Our use and interest of the term is taken from the work of the political philosopher Jacques Rancière (1999; 2013), who argues that this training tends to reinforce the status quo, rendering it more difficult to disagree with fundamental ways of organising and seeing the world. Making aesthetics political in this sense means addressing how the world is rendered communicable, and interrogating the conditions under which something new may present itself to the senses, or indeed, create a new public.

[3] One of the early theorists of complex systems theory, C.S. Holling, promoted the characterisation of the environment in terms of turbulent and unpredictable dynamics. The geographer Sarah Nelson (2015) argues convincingly that the emergence of 'resilience' as the key paradigm for managing complex ecosystems must be understood within the broader social, political and economic context of the 1970s and the rise of neoliberalism. Rather than seeing these dynamics as threats that had to be warded off, Holling argued for new 'resilient' forms of organisation that could 'harness the uncertainty generated by non-linear dynamics as a catalyst for innovation and growth' (Nelson 2015: 4).

[4] See the important feminist and decolonial work of Carolyn Merchant (1981), Deborah Bird Rose (2004), Val Plumwood (2002) and Silvia Federici (2004).

[5] President Roosevelt's important conservation advisor Gifford Pinchot was a delegate to the first and second International Eugenics Congress, in 1912 and 1921, and a member of the advisory council of the American Eugenics Society, from 1925 to 1935.

[6] Conservationism exploded during Theodore Roosevelt's administration (1901–1909). In his seven years as president, Roosevelt proposed the US Forest Service and increased the national forests by 148 million acres, founding five new national parks and over 50 wildlife refuges.

[7] In popular models for zoning Biosphere Reserves – designated areas of high biodiversity governed by the United Nations Educational, Scientific and Cultural Organisation – 'core' zones may be allotted for wildlife sanctuaries, while multiple-use zones permit some forms of ecologically sensitive extraction, as well as buffer zones. Zoning processes

8   can lead to conflict by imposing these areas on existing residents, but also, because each zone requires contradicting forms of 'protection'.

8   On the postcolonial geographies of *terra nullius*, see in particular the analysis of Makki (2014), Head (2000), Jazeel (2005), Millner (2016), and the edited collection of Jackson (2017), which expands on questions of the liveliness of the 'natural' through a posthuman lens. Such scholars emphasise how the colonial division of the world into territories for extraction not only works by classifying 'wild' frontiers against civilised centres, but by rendering landscapes legible in ways that produce bodies as inert, enabling extraction (Yusoff 2018).

9   Feminist and critical race scholars have extensively traced such systemic logics of violence and exclusion. See, for example, Yusoff (2018) on the similarities between racial logics and geology as they emerged in the 17th century, reducing Black bodies and the earth to inert matter available for manipulation and white man's profit. See Schiebinger (2004) for a full account.

10 This idea was captured most succinctly in the well-known memo from Laurence Summers, then chief executive officer of the World Bank, in 1991. Summers urged the World Bank to encourage the migration of dirty industries to 'Less-Developed Countries', an argument that rested on the 'impeccable logic that countries in Africa are vastly under-polluted' (cited in Pellow 2007: 9).

11 For more on 'green grabbing' see Amador and Millner (forthcoming); Fairhead et al (2012).

12 We can see this in our own work. In Ireland, for example, local opposition to large-scale, corporate-owned energy infrastructures is dismissed as NIMBYism, ignoring decades of underdevelopment in rural areas, the 'extractivist' gaze of the postcolonial state, and place-based demands to protect land-based livelihoods (Bresnihan and Brodie 2021).

13 Anthropologist Anna Tsing (2011: 202) remarks that 'Where indigenous people are recognised within [forests], international discussion makes the people wild things too, assimilated to the wild emptiness of forest.' Here she refers to the practices through which first forests, and second Indigenous groups, are made equivalent to 'wild things' and therefore made a backdrop against which environmental organisations and other actors can act as authorities to determine what needs to happen.

14 While social reproduction is most often associated with human reproduction and the management of the 'household', from childbirth, to childcare and healthcare, cleaning and cooking, it also extends beyond the confines of the house narrowly construed as four walls. Federici herself describes how her time in Nigeria observing and documenting the labour and activity of women in mostly subsistence economies led her to extend the notion of social reproduction (Federici 2018). Household (*oikos*) becomes not just a home or family but a wider sphere of communal reproduction that involves direct relations with the land, water, plants and animals.

15 It is also important that corporate environmental offenders like Chevron, DuPont, ExxonMobil and Monsanto can also be environmental donors in this framework. Dorsey (2005) has noted that the boards of directors of major conservation NGOs in the US are now dominated by the chief executive officers of major corporations, making it difficult for big NGOs (BINGOs) to take a hard stand on environmental issues that might contradict the interests of their corporate sponsors/board members. Environmental mitigation projects and eco-development corridors present conservation BINGOs with significant sources of funding, but often also constrain their agendas and voice.

16 For example, the four-yearly International Union for the Conservation of Nature World Conservation Congress, recent COP 2021, and also more entertainment oriented biannual Wildscreen natural history film festival in the UK, are orchestrated as spectacles where

corporate-conservation and media networks, as well as financialised understandings of nature and conservation interventions, are produced and reinforced (Sullivan 2013).

[17] In the US, the environmental justice movement grew out of an understanding that not only were toxic waste dumps and infrastructures detrimental to public health, but that the expert planning and science informing where they should be located was racist. In 1983, a US government study found three out of four hazardous waste landfills in the southern States were located in predominantly African American communities despite African Americans making up only 20 per cent of the regional population at the time (McGurty 1997).

[18] As the work of Maria Puig de la Bellacasa (2017; 2015; 2010) and others emphasise in relation to matters of soil, when such collective praxis is brought to bear on material horizons, we may see quite distinct kinds of temporality and ethos to those animating the dominant forms of agricultural production. The idea here is that not only forms of knowledge but modes of enacting care and even time (for example, slowness) have been marginalised by the progress-tempo of the Green Revolution, and must be 'repaired' as part of new kinds of knowledge politics.

[19] See http://www.authorityresearch.net/

Chapter 2

[1] This circulation and resonance was not, however, uniform. In other regions, contemporaneous movements relating to the risks of chemical pesticides articulated their claims through other motifs and via other thinkers. This meant that the reception of *Silent Spring* was itself geographically varied: the book was quickly translated into multiple European languages, but its impact in other places was more patchy, reflecting different agricultural systems, an unwillingness to accept that widespread chemical use was a 'European' problem, and a greater preoccupation with nuclear radiation (Stoll 2012).

[2] See for example Russell (2001) and Perkins (2012), both of which trace the chronology of pesticide development and contestation.

[3] Attention to these issues makes Carson's contribution an important intervention in science in feminist terms. Although, it has been noted that the labour undertaken by Carson's often female field assistants, some of whose observations were critical to her arguments, is rarely acknowledged in her published work.

[4] Mitchell reminds us of John Steinbeck's monumental rebuke to the Californian dream in 1939, 50 years before Starr wrote his epic trilogy on the history of California. It is testament to the power of ideology, the 'lie of the land', that Starr's imagined landscape, and the exploitation it hides and sustains, persists despite the well-received and celebrated work of a writer such as John Steinbeck.

[5] For more on the relationship between environmental health, citizen science and environmental justice see the work of Phil Brown et al (2011) on 'popular epidemiology'.

[6] In 1988, Dolores Huerta would be beaten by police as she demonstrated against President Bush's refusal to regulate pesticides despite farmworkers and their children continuing to die of cancer (Morain 1988).

[7] In contrast with Green Revolution programmes, which frame the peasant or *campesino* farmer as 'backward', agroecological programmes see peasant farmers as the locus of agricultural innovation and focus on transmitting techniques for testing and evaluating the effectiveness of specific techniques (Martínez-Torres and Rosset 2010; Wittman 2011). See also Chapter 4.

[8] It is important to understand that agroecology and popular education build on similar understandings about experience and experimentation, devised through histories of resistance praxis (see Millner 2016). Huerta and Chávez also participated in this movement and were influenced by it.

Chapter 3

1. See Edwards (2010), Gabrys (2016) and Yusoff (2017) on the importance of science and technologies in producing architectures of global world-space and a sense of planetary connectedness.
2. Indeed, so-called Third World representatives made it clear at the Stockholm meeting, as they would at every subsequent international meeting on the environment and development, that their primary concerns were the social and economic development of their nations – self-determination – not 'global' environmental problems framed by the imperial nations.
3. In the late 1960s, the Club of Rome, a global organisation made up of businessmen, government administrators, economists and statisticians, was set up by the Organisation for Economic Co-operation and Development to better understand and develop responses to what became known within those circles as the 'world problematique', a meta-problem that included poverty, environmental degradation, political apathy and militancy, unplanned urbanisation, alienation, macro-economic problems and the rejection of traditional social values and institutions (Schmelzer 2017). To make sense of these relationships the Club of Rome commissioned a study to be undertaken in MIT using the new computing power that was available and the most advanced computer modelling techniques. Spearheaded by Jay Forrester, the MIT team published their results in 1972 in a report entitled *The limits to growth*.
4. In the same way that our book challenges some of the 'storylines' of environmentalism, Kristin Ross's (2008) extensive archival work on the period makes clear that some of the retellings of these protests have pushed out of collective memory important parts of the struggle, while refocusing attention on others. Thus, we talk now of May 1968 but not its longer prehistory, or the important year that follows. This means we miss, for example, the birth of a new anti-progressivist agricultural movement in the early 1970s in the Larzac region, a movement 'that would manifest a distinct "afterlife" in the form of the egalitarian rural radicalism of the Confédération Paysanne, with its attacks on McDonald's and on genetically modified food' (Ross 2008: 9).
5. As early as May 1964, students organised in solidarity with the Sengalese fight for independence. Later they would protest in relation to struggles taking place in Mali, Haute Volta, Mauritania and Gabon (Slobodian 2008).
6. It is sometimes hard to keep track of the movements and influences of these young scholars. For example, Walter Rodney, the Guyanese historian and activist, undertook his university studies in Jamaica and London, before taking up a position as lecturer in the University of Dar es Salaam. He also had various visiting positions and roles in US and Canadian universities throughout the 1970s (Rodney 2018 [1972]).
7. For a short account of Sicco Mansholt's relationship to 'no-growth' policies in this period see Juan Martinez-Alier's (2014) post: ' "Growth below zero": In memory of Sicco Mansholt'.
8. Often dismissed as 'backward' and 'primitive', these critical engagements and experiments with industrial technology offer glimpses of possible futures that resist linear, progressivist accounts of social development. Gorz writes: '[I]t is not a question of reverting to cottage industry, to the village economy, or the Middle Ages, but of subordinating industrial technologies to the continuing extension of individual and collective autonomy, instead of subordinating this autonomy to the continuing extension of industrial technologies' (1980 [1975]: 34).
9. In a flyer from 1979, the Porto Maghera group states: 'Today we need to challenge the organisation of work, but not just its levels and parameters, but also the fundamental choices that make us produce commodities and not wealth, useless things made only to generate profits. ... *What, how, and how much to produce must be our parameters of struggle*' (cited in Feltrin and Sacchetto 2021: 14; emphasis in original).

10 Nearly a decade earlier, in 1970, the United Automobile Workers of America issued a similar statement to General Motors Corporation asserting that their members had a direct concern in pollution caused by the automobile industry, both within the manufacturing plants and through the emissions from combustion engines (Rector 2014).

11 These humanist engagements with technology links to the early work of Marx on alienation, as well as a long, radical tradition of experimental techno-politics traced from the Luddites, through William Morris and the Craft Movement, to the Soviet Proletkult (Wark 2015).

12 *Operaismo* (workerism) generally refers to a form of class-based politics which emerged during the 1960s in Italy. Based on a new reading of Marx by intellectuals such as Raniero Panzieri and Mario Tronti, *Operaismo* reflected a rejection of the reformist, parliamentary strategies of both the communist and socialist parties, focusing political organising around a new analysis of exploitation based on the worker's immediate experience in the workplace (Bracke 2013).

13 '[I]f we want humanity to advance a step further', Fanon writes, 'if we want to bring it up to a different level than that which Europe has shown us, then we must invent and we must make new discoveries' (Fanon 2001 [1963]: 254).

14 In his recent book, *Reconsidering reparations* (2022), Olúfẹ́mi O. Táíwò makes the case for a framework of reparations that is not primarily about financial payments, but about a 'construction project' that requires funds and resources to build autonomous institutions and communal forms of life. As we do in this chapter, Táíwò traces the origins of such a framework to the mid-20th-century decolonisation movement in the Third World.

15 In 1961, at the end of the Algerian War for Independence, Fanon identified land 'as a primary site of postcolonial recuperation, sustainability and dignity' (cited in DeLoughrey and Handley 2011: 3). In *The wretched of the earth* Fanon (2001 [1963]) calls upon his readers to revolt against imperialism in all its forms and create a new mode of relating to land as the most essential value. Fanon also brings us to the question of DDT in quite a different way, writing that values are 'irreversibly poisoned and infected as soon as they encounter the colonised. The customs of the colonised, their traditions, their myths, especially their myths, are the very mark of this innate depravity. Therefore, we should place DDT, which destroys parasites, carriers of disease, on the same level as Christianity, which roots out heresy, natural impulses, and evil. [The] triumphant reports by the missions in fact tell us how deep the seeds of alienation have been sown among the colonised' (2001 [1963]: 7). Here we read that the poison acting on the land is not only the chemical, but the logic that views certain ways of life as mal-adapted, extraneous and corrupting. Recalling the war metaphors used in the application of chemical pesticides and weapons discussed in Chapter 2, Fanon here connects it to a deeper and longer colonial practice of rooting out and destroying Indigenous cultures.

16 In 1956, Cabral co-founded the Partido Africano da Independencia da Guine e Cabo Verde and the Movimento Popular Libertação de Angola. Both these organisations were central to the struggle against (and ultimate victory over) Portuguese imperial rule in Africa.

17 The inspiration for Wynter's ideas were partly inspired from her early experiences of migrant labour (her family migrated to Cuba to seek work as sugar plantation-workers), labour struggles (in the Caribbean, these reached historic intensity during Wynter's youth, in the 1930s), anti-colonial social movements (during the 1940s and 1950s, there were unprecedented mass rebellions in the Caribbean by the region's poor against the colonial order) and her subsequent migration from the heartlands of colonial occupation into metropolitan London (Rodriguez 2015).

18 Asturias uses magical realism to highlight Mayan cultures and customs in their own terms. This is interesting because Asturias, like other Guatemalans educated abroad, was heavily

[19] critiqued by other emerging Indigenous writers in the same period, and his lyric exaltation of Indian and pre-Columbian myth formed a referent against which others positioned their work. His writing was seen by critics as being 'faddish' and a romanticisation of Indigenous culture fostered by living in colonial centres (Wainwright and Lund 2016).

[19] The Creole in Asturias's context refers to Spanish-descended colonisers, rather than the new peoples and languages that emerged in the Caribbean as peoples of Carib, European, Indigenous and African-descent backgrounds intermingled.

[20] Today, André Gorz is closely associated with the political project of degrowth, best defined as a democratic process enabling global social and ecological justice, popular self-determination, and the redesign of institutions and infrastructures away from the imperative for continuous growth (Schmelzer et al 2022).

Second Interlude

[1] The work has also been attributed to an earlier author, Zayn al-Din Abu Hafs 'Umar Ibn al-Muzaffar Ibn al-Wardi, who died in 1348 AD, but according to the 1479 and 1487 copies of the Kharidat, this author would pre-date the original work.

Chapter 4

[1] Unprecedented in its scale of reporting and data collection from participating nations, the Report identified important challenges, but unlike predecessors like the Club of Rome's *Limits to growth* (Meadows et al 1972), its rhetoric suggested that these challenges could be overcome through institutional and technological innovations. In this new way of thinking, environmental problems offered economic and technological opportunities, particularly to stagnating former industrial nations in North America and Europe.

[2] In 1992, the American political scientist Francis Fukuyama (2006 [1992]) published *The end of history and the last man*, the extension of an essay he had initially published in 1989. His thesis: the collapse of the Soviet Union was not just the end of an era of communism, but the end of history itself, understood through German philosopher Hegel as the linear unfolding of a universal culture. Western, liberal democracy and capitalist economic systems were, he suggested, the ultimate destination-point.

[3] The Midnight Notes Collective was an extension of the Zerowork collective discussed in Chapter 3 (that is, individuals involved in anti-colonial struggles, Italian autonomist Marxism and the Women's Liberation Movement in the 1960s/1970s). It is not surprising, then, that their analysis would have been quick to speak to the global shifts in capitalist development in this period, particularly regarding questions of social reproduction and antagonism beyond the wage relation.

[4] Structural Adjustment Programmes required often newly independent nation states across the Global South to open up territories, resources and labour to the extractive and exploitative relationships of the global market, primarily via bilateral and multilateral free trade agreements (McNally 2006). Such agreements made privatisation, deregulation and economic restructuring effectively mandatory for nation-states in the Global South if they wished to remain within debt assistance programmes. As many political ecologists, geographers and social scientists have pointed out, these policies were terraforming the planet in destructive ways, dispossessing populations in the process (Esteva and Prakash 1992; Escobar 1995).

[5] The special issue was written by over a dozen authors located in Africa, North America, Asia, Europe and Australia.

[6] This general and multifaceted process of marketisation in the area of environmental governance has led some scholars to describe it as the 'neoliberalization of nature' (Mansfield 2004; Heynen et al 2007).

[7] Scholars have called this 'green governmentality', building on Foucault's (2007) theoretical architecture. Where governmentality refers to the art of governing subjects as part of state rationality – including ideas of the population, security practices and notions of political economy – that has been evolving (in Europe specifically) since the 1750s, green governmentality refers to the way that this has been configured in relation to the environment (Reid 2013). Green governmentality is characterised through notions of the stewardship of nature and management of its resources, and with the emergence of a new set of administrative truths that organise and legitimise common understandings of an 'environmental' reality as well as how things should be between humans and nature (Bresnihan 2016; 2020).

[8] The commitment to community-led natural resource management initiatives was articulated in sustainable development theories as early as the 1970s (Fairhead et al 2012), as a result of critiques of the paternalistic premise of development interventions, and fresh conservation commitments in the face of global biodiversity loss. Participatory approaches aimed to correct imbalances in previous decades of top-down development by involving lay actors and marginalised communities in decision-making and were scaled into international commitments at the 1992 Earth Summit. However, the reconfiguration of responsibilities associated with the rise of participatory models also reflects a shift from 'government' to 'governance', where civil society assumes responsibility for achieving outcomes determined at state and transnational levels (Jessop 2002; Millner et al 2020).

[9] In the 1980s, the concept of the Brandt Line was developed as a way of showing how the world was geographically split into relatively richer and poorer nations. This was a line that could be drawn on the world map to indicate how countries in the northern hemisphere had tended to accumulate global wealth, whereas those in the south had remained indebted and 'underdeveloped'. Although somewhat simplistic, this term was adopted by the Non-Aligned Movement to refer to economically disadvantaged nation-states and as a post-Cold War alternative to 'Third World' (Mahler 2018).

[10] As we suggested in Chapter 2, agroecology, since its inception has formed an important, science-based but lay movement that shares many concerns with modern environmentalism (for example, modern pesticides), yet has retained affinities with labour, anti-colonial and Third-Worldist/Global South led forms of organising (see Millner 2017). Although it spread initially through the support of INGOs the movement has always foregrounded local leadership and has been characterised by 'bottom-up' forms of organising. Across Central America, the Campesino-a-Campesino (farmer-to-farmer) model (Holt-Giménez 2006) grew rapidly during the late 1980s when heavy flooding exposed the difference between traditionally terraced farms, and modern farms, which were stripped of topsoil.

[11] In Mayan portrayals of soil life, both what has been lost (land, cultural practices, soil fertility) and what is to be restored, are joined together. The Yucatec Maya term for soil is *Lu'um*, which also means land, terrain, landscape and nature – as well as life-supporter, home, territory, womb and graveyard (Barrera-Bassols and Toledo 2005). *Lu'um* is a comprehensive relational domain of life, while *Santo Lu'um*, the 'spirit of the land' is one of the most important deities. Soil is also associated with the capacity to heal and restore.

[12] Agriculture in Mesoamerica is already considered syncretic, for agricultural practices grounded in Indigenous modes of seeing the world (including soil management, terracing, composting) persist even in areas where Euro-western, 'modern' techniques have drastically altered the landscape.

[13] Of course, others, including feminist and ecofeminist thinkers, have long explored the capacities of earth mother figures within wider environmental critiques, as well as their limitations. Ecofeminist movements have connected the struggles of environmentalism and feminism in important ways, drawing connections between the systematic violence experienced by women (especially women of colour) and the systematic violence wrought

on the earth through extractivism and masculinist science (Merchant 1981; Plumwood 1986). Yet such movements also created contexts for the re-evaluation of the same figures used to connect across these struggles. Recent work in black womanist, Chicanx and Latinx studies also takes back concepts of woman and mother in important ways, disrupting key assumptions that underpin specifically white feminist and ecofeminist thought (Millner 2021).

14 The invitation to 'queer' nature is to question the normative uses to which it is set, and the 'relations of knowledge and power by which certain truths about ourselves have been allowed to pass without question' (Sandilands 1997: 35). Gloria Anzaldúa (1987), for example, revisits the genealogy of Guadelupe, the Mexican virgin Mary figure, to emphasise that her Mexican name contains remnants of earth mother figures. Rather than abandon her to Catholic canon she wants this figure reclaimed and included as an alternative to the virgin versus whore dichotomy offered to Mexican and Chicana women as options for femininity.

15 This explicitly postcolonial genre made its way to the forefront of cultural production in Guatemala during the 1970s when the agribusiness sector began to expand into uncontested Indigenous lands with strong backing from its US-supported military (Beverley and Zimmerman 1990). In response, Indigenous communities began to resist in armed and nonviolent ways, and to devise narrative forms that could communicate the accompanying assassinations, disappearances and massacres to an international audience.

16 The Federal Republic of Nigeria became an independent nation in 1960, but it carried with it the institutional and economic imprint of British colonialism. Divided into 36 states, the country has over 250 ethnic groups, speaking 500 distinct languages. As with other British colonies, the use of 'divide and rule' as a strategy of colonial administration relied on the fomenting of cultural and ethnic differences, and the uneven distribution of power, resources and wealth between distinct groups. These hierarchies and distinctions have been the source of conflict and violence both during and after colonial rule. The Niger Delta in the South of the country is one of the principal sites through which this topology of power has operated.

17 The Ogonis live in the Rivers State district of the Niger Delta in a 1,000-square-kilometre (404-square-mile) territory which they refer to as Ogoniland (UNPO 2017). Rob Nixon (1996) refers to them as a 'micro-minority': in the 1990s, there were 500,000 Ogoni in a nation of some 140 million, composed of nearly 300 ethnic groups.

18 Anna Tsing (2011) deciphers something similar in the role that environmental discourses and practices played in radical Indonesian social movements throughout the 1980s and 1990s. Indonesian students inspired by sources as eclectic and surprising as Ralph Waldo Emerson forged attachments to landscapes under threat from industrial-scale logging, which became one force in an unlikely alliance of social and cultural groups opposing the Indonesian government and the intrusion of multinational corporations.

19 In 1991, an addendum was added to the Ogoni Bill of Rights entitled, 'An Appeal to the International Community'. It concluded: '[w]ithout the intervention of the international community, the government of the federal Republic of Nigeria and the ethnic majority will continue these noxious policies until the Ogoni people are obliterated from the face of the earth' (Saro-Wiwa 2012 [1995]: 63).

20 The work of postcolonial writers such as Guha (1989) and Said (1978) are vital to Nixon's argument that this is a form of violence that has a specifically colonial history: it is allowed to happen in some places and to some humans, rendered 'disposable', while others remain protected and at a remove from even hearing about the effects.

21 Whether or not Saro-Wiwa explicitly mentioned the boycott of South African goods in the late 1980s, this was a powerful international campaign that strengthened the African National Congress and popular movements against Apartheid.

[22] The same year, Saro-Wiwa travelled to Geneva to speak to the UN Working Group on Indigenous Populations and Unrepresented Nations and People's Organisation (UNPO) – representatives in UNPO included Tibet, Chechnya, Abkhazia and Bougainville. As well as being the year of the Earth Summit, 1992 was the UN International Year for the World's Indigenous populations, with an event held in New York.

[23] It is hard not to think of Roger Casement's advocacy work on behalf of the people of Congo in the early part of the century, compiling evidence and providing accounts of Belgian atrocities for an international audience.

[24] In 2005, Ken Saro-Wiwa's brother, Dr Owens Wiwa, attended the trial of five men who had been imprisoned for refusing to obey a court injunction forbidding them to interfere with work being undertaken by Shell on their land. Dr Wiwa said: 'their courage gives hope to the African people … to see that Shell can be made to back down … it is being watched with interest by many involved in similar situations' (quoted in Bowman 2008).

[25] The year before the Zapatista insurrection, the Mexican president, Salinas, under pressure to attract foreign investment and secure NAFTA, altered Article 27 of the Mexican Constitution to allow for the privatisation of communal land. Article 27 was one of the last meaningful guarantees of indigenous and community integrity in the Chiapas. It was this that sparked the Zapatista communities into ordering their citizen army, the EZLN, to take action.

[26] When Ruiz returned to Mexico, he went to the villages in the Chiapas on foot, and just as Marcos two decades later, he was forced to admit: 'I came to San Cristobal to convert the poor, but they ended up converting me' (cited in Marcos 2011: 294).

[27] Five hundred years of colonialism and extractive industries comprise the 'wind from above', which is rivalled by the 'wind from below', the subsistence commons, but also the ghosts of unfinished histories (Marcos 1992).

[28] Harry Cleaver Jr., writing in 1998, describes how the Zapatista inter-continentals meeting held in Germany and Canada in 1996 involved activists discussing at length the similarities and differences between Latin American 'Neoliberalism', the 'Thatcherisation' of the economy, the move towards greater European integration found in the Maastricht Treaty, and the US–Canadian experiences with Reagan's supply-side economics (Cleaver 1998). 'The result of such deliberations', Cleaver writes, 'was a commitment to collective and coordinated opposition to what is perceived as homogeneous global policy' (Cleaver 1998: 631).

[29] Other important figures in this decade include the biologist Wangari Maathai, who claimed that national debt to 'first world' agencies such as the World Bank and the International Monetary Foundation are one of biggest obstacles to environmental sustainability (Clapp and Dauvergne 2011).

Third Interlude

[1] CODECA organises campaigns and demonstrations across Guatemala in relation to human rights abuses of those trying to protect the environment. They also work to improve the situation of the rural poor in Guatemala, focusing on issues such as the wage conditions for farmers, land reform and nationalisation of electric energy in the country. Here we refer to the regional branch of CODECA, which operates across the Petén region of Guatemala.

[2] On 11 July 2012, human rights defender Enrique Linares, a member of CODECA, was shot dead in Zacapa in Guatemala. Furthermore, in the January before this demonstration took place, Antonio Cruz Jimenez, community defender of human rights from Vancia Jutiapa, had been hit by a truck on his way to a demonstration the previous week. Cruz Jimenez was the leader of the central core of the organisation.

3   The Q'eqchi', whose land was given away to European colonists, had long been framed as 'less indigenous', because they lived on lowland plantations owned by colonists, as opposed to the Maya-K'iche and Kaqchikel of Guatemala's highlands, more likely to speak Spanish, who were consistently favoured by military planners. As international conservation schemes were established in the neighbouring Petén, Q'eqchi' agricultural practices, involving mixed swiddens with long fallow periods, were framed by international actors like the World Bank as 'nomadic squatting' instead of knowledgeable stewardship – and as potential threats to conservation practice (Ybarra 2018: 36, 40).

Chapter 5

1   In the introduction to her book (2020), Barca also takes the 'Welcome to the Anthropocene' film as a jumping off point for her critique of the 'official' Anthropocene narrative. Closely aligned with our position, Barca shows that the film's narrative excludes the agency of what she calls 'reproductive forces': not just women's bodies and historically feminised and racialised care work, but all struggles for social and ecological reproduction more broadly, led by environmental justice, peasant and other commons-based movements. Barca includes in this category, the historical agency of waged workers' struggles for health and safety and for the regulation of hazards and toxicity in industrial production (see Chapter 2).
2   While Earth Systems Science emerged in the post-war era, it wasn't until the 1980s, in response to demands for a new 'science of the Earth', that it began to develop in earnest, becoming today the dominant paradigm for understanding global environmental change.
3   The work of critical geographers on the emergence of natural capital and ecosystem services has been particularly important in this regard (Robertson 2006; Sullivan 2018a). For example, Sian Sullivan (2018b) discusses the shift towards 'Nature 3.0' as new blockchain technologies prompt the advance of supposedly frictionless commodification and exchange of natural assets. She explores this through an analysis of the Natural Asset Exchange, an emerging online platform established by Mauritius-based company impactChoice. Its aim is to offer efficient, transparent and democratic connections between producers, buyers and consumers of so-called natural capital assets.
4   It would later lead to the production of *The Universal Declaration of the Rights of Mother Earth* that was used to articulate the rights of nature for the first time in Bolivia and Ecuador's state law regimes (Caputi 2016). What is significant about this manifesto, in ontological terms, is the way it attempts to translate an alternative fabric of knowing life into the heart of modern institutions. Some see this effort as a failure: in Bolivia, it did not disrupt the elaboration of the neo-extractive imperative and the suppression of dissenting Indigenous voices (Gudynas 2013). When the Indigenous idea of *Sumak Kawsay* was translated into policies for '*buen vivir*' (living well), there were likewise objections that what was going by the name of *buen vivir* did not equate in any way to Indigenous modes of coexisting with the more-than-human world (Gudynas 2011; Radcliffe 2012). What is evident through such failures, however, is that *Madre Tierra* centres not on an alternative set of environmental policies, but on a dispute between non-compatible modes of *worlding* the world.
5   ALBA includes as its member nations the founding countries Cuba and Venezuela (2004), together with Bolivia (2006), Nicaragua (2007), Dominica (2008), Ecuador (2009), Saint Vincent and the Grenadines (2009), Antigua and Barbuda (2009), and Haiti and Uruguay as 'observer states' (non-official members included in regional conferences).
6   This analysis clarifies the nature of the *politics of aesthetics* that is at stake in contemporary environmental politics. Political philosopher Jacques Rancière names this as a problem of the 'distribution of the sensible' (Panagia 2014; Rancière 2015). This means that it matters how concerns are communicated, because the figures we tend to rely on frequently serve to reproduce dominant ways of seeing and saying the world, rather than disruptively *re*-figuring the world, as we may hope.

[7] Until this takes place, the language of translation remains critical here. For example, to save the earth-being Ausaungate from the threat of mining exploitation, the Runakuna people had to stop focusing on the spiritual aspects of *tirakuna* in their negotiations and instead transform their politics into the more familiar terms of 'environmental' defence. This, however, was not a true recognition of Ausaungate in their terms, but a settlement that reduced the earth being to 'cultural' and 'environmental' attributes.

[8] Escobar does also trouble the legal recognition of nature as a political actor, however, noting that recognising the voice of nature in legal frameworks may end up silencing 'earth-beings' by requiring them to make their claims in juridical terms, to submit to the policing logic of the nation-state, and to occupy only one side of the nature–culture binary that they offer to unsettle. If the constitutionalisation of nature is to be effective in building an identity politics, its simplification of the terms in play may prove detrimental to the complexities of Indigenous thought and expression.

[9] Permaculture very much emerged on the edge of environmental paradigms and as a corrective to mainstreaming sustainable development paradigms. Like agroecology, it took issue with, and sought to enact disagreement with, top-down or market-based solutions to food poverty, agriculture, and so on, and was an early adopter of understanding the importance of the dependence of different species in forest systems, rather than relying on single crops in productive agriculture.

[10] The process is credited to Professor Higa, a Japanese horticulturist and agricultural scientist based at the University of the Ryukyus, Okinawa, who sought alternatives to the accelerating use of chemical fertilisers during the 1970s, when the Green Revolution reached Japan (Higa and Parr 1994).

[11] Today, for example, wind is championed as the energetic force that can sustain life as we know it after the end of fossil fuels; wind-as-energy evokes dominant Anthropocene imaginaries of technological mastery and abundant frontiers (Bresnihan 2022).

[12] Activists at Standing Rock preferred to designate themselves as 'water protectors'. They regarded the term 'protester' as a colonised term used by those in power to categorise and delegitimise those standing up for rights and values. The term water protector is rooted in Indian indigenous perspectives on the duty of humans to respect and honour water.

[13] The popular hashtag #Mniwiconi, for example, comes from a Lakota term, *Mní Wichóni*, which is usually translated as water is life. However, the translation loses many of the subtleties of meaning involved in the term when removed from its context. 'Life' here carries a series of assumptions about the interconnectedness of life and the obligations and sacrifices involved in protecting life – 'Wichoni is actually larger than life. It is our resistance, our future generations, how we do things. We pray to the water and pray with it and pray for it. It's for real a relative. It's a relative that nourishes us' (Jewett 2017).

[14] Similarly, Dina Gilio-Whitaker points to the often-overlooked educational work that American Indian activists engage in when doing environmental justice organising – education directed at those, particularly in environmental organisations and groups, who know little about the histories that have deliberately been marginalised and excluded. Educating groups and people with whom they may have contentious relations, 'points to an in-between space in Indigenous environmental justice organizing', she writes, 'where provisional alliances and new articulations of decolonial environmental justice concerns might be composed' (Gilio-Whitaker 2019: 10).

[15] In 2009, three years before Cochabamba, the Indigenous Peoples' Global Summit on Climate Change was held in Anchorage, Alaska. Indigenous representatives from the Arctic, North America, Asia, Pacific, Latin America, Africa, Caribbean and Russia gathered on Ahtna and Dena'ina Athabascan Peoples land. The Anchorage Declaration called for 'states to return and restore lands, territories, waters, forests, oceans, sea ice and sacred sites that have been taken from Indigenous Peoples, limiting our access to our

traditional ways of living, thereby causing us to misuse and expose our lands to activities and conditions that contribute to climate change' (Anchorage Declaration 2009: 3–4).

[16] Maria Puig de la Bellacasa's work on soil and 'care time' is instructive here, looping ideas of repair and continuity with posthumanist ontologies and technoscientific practices (2015). 'Making time for soil' involves different technoscientific practices (not their abandonment) that object to the short-term, productivist temporalities of soil management, expanding possibilities for relations of care to thicken in the present and thus open up alternate futures.

[17] In his book, *Experimental practices* (2018), Papadopoulos describes an historically, geographically and culturally eclectic series of 'more-than-social movements' that have developed through such decolonial practices of matter, that is, making new worlds rather than seeking their representation or inclusion within existing governance arrangements. This includes ecological movements relating to the reclaiming of land, alternative forms of agriculture, seed saving, as well as HIV activism that reconfigured relations to the body and health through collective infrastructures of care.

[18] From the perspective of Indigenous peoples, and non-Indigenous acting in solidarity with them, these two demands are necessary parts of decolonisation: the immediate and long-term effects of continued fossil fuel extraction and combustion undermine Indigenous sovereignty, as do various legal, political and economic structures.

[19] Chas Jewett is returning to Ireland in summer 2022 with 21 other activists from Standing Rock, this time to show solidarity with communities across Ireland opposing mining (Cirefice 2021). Part of their visit will involve a four-day water walk from Greencastle to Derry, bringing together different groups across the Irish border.

Fourth Interlude

[1] http://www.mariatherezaalves.org/works/seeds-of-change-a-floating-ballast-seed-garden-bristol?c=
[2] https://civiclaboratory.nl/
[3] https://deirdre-omahony.ie/portfolio/t-u-r-f/
[4] https://www.climavore.org/

Chapter 6

[1] There are many but two examples of recent projects that have begun in Europe are the UK-based 'Land In Our Names' (https://landinournames.community/) which connects land justice with issues around race, food insecurity, health inequalities and environmental injustice, and 'Common Ecologies' (http://commonecologies.net/), working to build transversal movements through struggles relating to social reproduction and ecology.

[2] FA works with partners across civil society to carry out investigations with and on behalf of communities affected by conflict, police brutality, border regimes and environmental violence (Forensic Architecture 2021).

[3] The collaboration aims to locate plantation cemeteries historically omitted from land surveys and maps in the region. By locating the cemeteries they may not only block planning permission for the new plastics factory, but also connect their larger struggle for ecological reparations and environmental justice with those of their Black descendants and the violent environmental histories of Louisiana. As with other counter-mapping efforts, this is a project of *anti-erasure* (see kollektiv orangotango+ 2018).

[4] Rodrigo Nunes' (2021) proposes a way to think of organisation ecologically which captures something similar to our idea of resonance. He writes: 'there need be no kind of coordination or even direct contact among the different components of an ecology for them to interact with one another: by acting on their shared environment, they can indirectly shape each other's fields of possibilities' (2021: 211).

References

Agard-Jones, V. (2014). Spray. *Somatosphere*, 27 May. Available at: http://somatosphere.net/2014/spray.html/ (last accessed 24 May 2022).

Agarwal, A. and Narain, S. (1991). *Global warming in an unequal world: A case of environmental colonialism*. Available at: https://www.osti.gov/etdeweb/biblio/6842576 (last accessed 16 May 2022).

Agrawal, A. (1995). Dismantling the divide between indigenous and scientific knowledge. *Development and Change*, 26(3), 413–439.

Agrawal, A. and Sivaramakrishnan, K. (2000). *Agrarian environments: Resources, representations, and rule in India*. Duke University Press.

Ajl, M. (2021). *A people's green new deal*. Pluto Press.

Alaimo, S. (1994) Cyborg and ecofeminist interventions: Challenges for an environmental feminism. *Feminist Studies*, 20(1), 133–152.

Allen, W. (2008). *The war on bugs*. Chelsea Green Publishing.

Altman, R. (2019). Time-bombing the future. *Aeon*, 2 January. Available at: https://aeon.co/essays/how-20th-century-synthetics-altered-the-very-fabric-of-us-all (last accessed 24 May 2022).

Amador, M. and Millner, N. (forthcoming). Being paramuno: Peasant world-making practices in the paramos (high moorlands) of the Colombian Andes. *Society & Natural Resources*.

Anchorage Declaration (2009). Indigenous Peoples' Global Summit on Climate Change. Available at: https://unfccc.int/resource/docs/2009/smsn/ngo/168.pdf (last accessed 14 February 2022).

Anker, P. (2005). The ecological colonization of space. *Environmental History*, 10(2), 239–268.

Antink, B. (2020). Bristol's Colston statue was toppled because racist historical narratives have not been effectively challenged. *RSA Online*. Available at: https://www.thersa.org/blog/2020/06/bristol-colston-statue?gclid=CjwKCAiAsNKQBhAPEiwAB-I5zfQMWYyPURmFeY8suzV71RPNHp_-j4FDDlK-fiw4VHE-9PIUhjyl_RoCtIUQAvD_BwE (last accessed 22 February 2022).

Anzaldúa, G. (1987). *Borderlands/la frontera: The new mestiza*. Aunt Lute Books.

Ari, W. (2014). *Earth politics: Religion, decolonization, and Bolivia's indigenous intellectuals*. Duke University Press.

Arsel, M. and Büscher, B. (2012). Nature™ Inc.: Changes and continuities in neoliberal conservation and market-based environmental policy. *Development and Change*, 43(1), 53–78.

Astier, M., Argueta, J.Q., Orozco-Ramírez, Q., González, M.V., Morales, J., Gerritsen, P.R.W., et al (2017). Back to the roots: Understanding current agroecological movement, science, and practice in Mexico. *Agroecology and Sustainable Food Systems*, 41(3–4), 329–348.

Bäckstrand, K. and Lövbrand, E. (2006). Planting trees to mitigate climate change: Contested discourses of ecological modernization, green governmentality and civic environmentalism. *Global Environmental Politics*, 6(1), 50–75.

Barca, S. (2019). Labour and the ecological crisis: The eco-modernist dilemma in western Marxism(s) (1970s–2000s). *Geoforum*, 98, 226–235.

Barca, S. (2020). *Forces of reproduction: Notes for a counter-hegemonic anthropocene.* Cambridge University Press.

Barca, S. and Leonardi, E. (2018). Working-class ecology and union politics: A conceptual topology. *Globalizations*, 15(4), 487–503.

Barrera-Bassols, N. and Toledo, V.M. (2005). Ethnoecology of the Yucatec Maya: Symbolism, knowledge and management of natural resources. *Journal of Latin American Geography*, 4(1), 9–41.

Bebbington, A. and Carney, J. (1990). Geography in the International Agricultural Research Centers: Theoretical and practical concerns. *Annals of the Association of American Geographers*, 80(1), 34–48.

Beckett, C. and Keeling, A. (2019). Rethinking remediation: Mine reclamation, environmental justice, and relations of care. *Local Environment*, 24(3), 216–230.

Bennholdt-Thomsen, V. and Mies, M. (2000). *The subsistence perspective.* Zed Books.

Beverley, J. (1989). The margin at the center: On *testimonio* (testimonial narrative). *Modern Fiction Studies*, 35(1), 11–28.

Beverley, J. and Zimmerman, M. (1990). *Literature and politics in the Central American revolutions.* University of Texas Press.

Bhandar, B. (2018). *Colonial lives of property: Law, land, and racial regimes of ownership.* Duke University Press.

Biermann, C. and Mansfield, B. (2014). Biodiversity, purity, and death: Conservation biology as biopolitics. *Environment and Planning D: Society and Space*, 32(2), 257–273.

Blaikie, P. (2016 [1985]). *The political economy of soil erosion in developing countries.* Routledge.

Bonneuil, C. and Fressoz, J. (2016). *The shock of the Anthropocene: The earth, history and us.* Verso Books.

Bookchin, M. (1974 [1962]). *Our synthetic environment*, vol 363. Harper & Row.

Borras, S.M. (2010). The politics of transnational agrarian movements. *Development and Change*, 41(5), 771–803.

Boudia, S. and Jas, N. (eds) (2014). *Powerless science? Science and politics in a toxic world*, vol 2. Berghahn Books.

Bowman, A. (2008). Shell to sea. *Red Pepper*. Available at: https://www.redpepper.org.uk/Shell-to-Sea/ (last accessed 15 June 2021).

Bracke, M.A. (2013). Between the transnational and the local: Mapping the trajectories and contexts of the Wages for Housework campaign in 1970s Italian feminism. *Women's History Review*, 22(4), 625–642.

Braun, B. (2002). *The intemperate rainforest: Nature, culture, and power on Canada's west coast*. University of Minnesota Press.

Brenner, N., Peck, J. and Theodore, N. (2010). After neoliberalization? *Globalizations*, 7(3), 327–345.

Bresnihan, P. (2016a). The more-than-human commons: From commons to commoning. In Kirwan, S., Dawney, L. and Brigstocke, J. (eds) *Space, power and the commons*. Routledge, pp 105–124.

Bresnihan, P. (2016b). *Transforming the fisheries: Neoliberalism, nature, and the commons*. University of Nebraska Press.

Bresnihan, P. (2019). Water, our relative: Trauma, healing and hydropolitics. *Community Development Journal*, 54(1), 22–41.

Bresnihan, P. (2020). Beyond the limits to growth: Neoliberal natures and the green economy. In Legun, K., Keller, J., Bell, M. and Carolan, M. (eds) *The Cambridge handbook of environmental sociology*, vol 2. Cambridge University Press, pp 124–142.

Bresnihan, P. (2022). Tilting at windmills. In Papadopoulos, D., de la Bellacasa, M.P. and Myers, N. (eds) *Reactivating elements: Chemistry, ecology, practice*. Duke University Press, pp 151–175.

Bresnihan, P. and Brodie, P. (2021). New extractive frontiers in Ireland and the moebius strip of wind/data. *Environment and Planning E: Nature and Space*, 4(4), 1645–1664.

Bresnihan, P. and Millner, N. (2022). Decolonising environmental politics. In Pellizzoni, L., Leonardi, E. and Asara, V. (eds) *Handbook of critical environmental politics*. Edward Elgar, pp 521–539.

Brigstocke, J., Bresnihan, P., Dawney, L. and Millner, N. (2021). Geographies of authority. *Progress in Human Geography*, 45(6), 1356–1378.

Brown, N., Griffis, R., Hamilton, K., Irish, S. and Kanouse, S. (2007). What makes justice spatial? What makes spaces just? Three interviews on the concept of spatial justice. *Critical Planning*, 14(6), 7–28.

Brown, P., Morello-Frosch, R. and Zavestoski, S. (eds) (2011). *Contested illnesses: Citizens, science, and health social movements*. University of California Press.

Brundtland Commission (1987). *World Commission on Environment and Development: Our common future*. Oxford University Press.

Buell, F. (2004). *From apocalypse to way of life: Environmental crisis in the American century*. Routledge.

Bullard, R.D. (2018). *Dumping in Dixie: Race, class, and environmental quality*. Routledge.

Cabral, A. (2009 [1970]). National liberation and culture. *Black Past*, 10 August 2009. Available at: https://www.blackpast.org/global-african-history/1970-amilcar-cabral-national-liberation-and-culture/ (last accessed 8 June 2022).

Cabral, A. (2016). *Resistance and decolonization*. Rowman & Littlefield.

Caputi, J. (2016). Mother earth meets the Anthropocene. In *Systemic crisis of global climate change: Intersections of race, class and gender*. Routledge, pp 20–33.

Carolan, M. (2013). *The real cost of cheap food*. Routledge.

Carson, R. (2002 [1962]). *Silent Spring*. Houghton Mifflin Harcourt.

Carson, R. (2018). *Rachel Carson: Silent Spring & other environmental writings*. The Library of America.

Ceceña, A.E. (2004). The subversion of historical knowledge of the struggle: Zapatistas in the 21st century. *Antipode*, 36(3), 361–370.

César, F. (2018). Meteorisations: Reading Amílcar Cabral's agronomy of liberation. *Third Text*, 32(2–3), 254–272.

Chabal, P. (1983). *Amílcar Cabral: Revolutionary leadership and people's war*, vol 37. Cambridge University Press.

Chakrabarty, D. (2009). *Provincializing Europe*. Princeton University Press.

Christian, D. and Wong, R. (eds) (2017). *Downstream: Reimagining water*. Wilfrid Laurier University Press.

Cirefice, V. (2021). The women keeping Ireland's gold in the ground. *The Ecologist*, July. Available at: https://theecologist.org/2021/jun/10/women-who-keep-gold-irelands-ground (last accessed 21 February 2022).

Clapp, J. and Dauvergne, P. (2011). *Paths to a green world: The political economy of the global environment*. MIT Press.

CLEAR (nd). Civic Laboratory for Environmental Action Research. Available at: https://civiclaboratory.nl (last accessed 19 July 2022).

Cleaver Jr, H.M. (nd). Background: Genesis of Zerowork #1. Available at: http://zerowork.org/GenesisZ1.html (last accessed 13 January 2022).

Cleaver Jr, H.M. (1998). The Zapatista effect: The internet and the rise of an alternative political fabric. *Journal of International Affairs*, 51(2), 621–640.

Connolly, J. (2013). *Songs of freedom: The James Connolly songbook*. Edited by M. Callahan. PM Press.

Connolly, W.E. (2007). Wolin, superpower, and Christianity. *Theory & Event*, 10(1), np.

Connolly, W.E. (2008). The evangelical-capitalist resonance machine. In *Capitalism and Christianity, American style*. Duke University Press, pp 39–68.

Cosgrove, D.E. (2001). *Apollo's eye: A cartographic genealogy of the earth in the western imagination*. Johns Hopkins University Press.

Coulthard, G. (2010). Place against empire: Understanding Indigenous anti-colonialism. *Affinities: A Journal of Radical Theory, Culture, and Action*. Available at: https://ojs.library.queensu.ca/index.php/affinities/article/view/6141 (last accessed 24 March 2023).

Cronon, W. (1996a). The trouble with wilderness: Or, getting back to the wrong nature. *Environmental History*, 1(1), 7–28.

Cronon, W. (ed) (1996b). *Uncommon ground: Rethinking the human place in nature*. W.W. Norton & Company.

Cullather, N. (2004). Miracles of modernization: The Green Revolution and the apotheosis of technology. *Diplomatic History*, 28(2), 227–254.

Dalla Costa, M. ([1972] 2019). Women and the subversion of the community. In Barbagallo, C. (ed) *Women and the subversion of the community: A Mariarosa Dalla Costa reader*. PM Press, pp 65–132.

Davies, A. (2022). *A world without hunger: Josué de Castro and the history of geography*. Liverpool University Press.

Davis, H. and Todd, Z. (2017). On the importance of a date, or decolonizing the Anthropocene. *ACME*, 16(4), 761–780.

De Angelis, M. (2011). Climate change, mother earth and the commons: Reflections on El Cumbre. *Development*, 54(2), 183–189.

De Chiro, G. (1996). Nature as community: The convergence of environment and social justice. In Cronon, W. (ed) *Uncommon ground: Rethinking the human place in nature*. W.W. Norton & Company, pp 298–321.

De la Cadena, M. (2010). Indigenous cosmopolitics in the Andes: Conceptual reflections beyond 'politics'. *Cultural Anthropology*, 25(2), 334–370.

DeLoughrey, E. (2011). Yam, roots, and rot: Allegories of the provision grounds. *Small Axe*, 15(1), 58–75.

DeLoughrey, E. and Handley, G.B. (2011). Introduction: Toward an aesthetics of the earth. In *Postcolonial ecologies: Literatures of the environment*. Oxford University Press, pp 3–39.

DeLoughrey, E.M., Gosson, R.K. and Handley, G.B. (eds) (2005). *Caribbean literature and the environment: Between nature and culture*. University of Virginia Press.

Demos, T.J. (2013). Contemporary art and the politics of ecology: An introduction. *Third Text*, 27(1), 1–9.

Desmarais, A. (2007). La Via Campesina: Globalization and the power of peasants. *Journal of Rural Studies*, 21(3), 359–371.

De Sousa Santos, B. (2005). The future of the World Social Forum: The work of translation. *Development*, 48(2), 15–22.

Dhada, M. (1993). *Warriors at work: How Guinea was really set free*. University Press of Colorado.

Dinkel, J. (2018). *The Non-Aligned Movement: Genesis, organization and politics (1927–1992)*. Brill.

Dorsey, M.K. (2005). *Commercialization of biodiversity: Processes, actors, and contestation in Ecuador, 1536–2001.* University of Michigan Press.

Dunlap, A. (2019). *Renewing destruction: Wind energy development, conflict and resistance in a Latin American context.* Rowman & Littlefield.

Ecologist, The (1992). Whose common future? Special issue. *The Ecologist,* 22(4).

Eddens, A. (2019). White science and indigenous maize: The racial logics of the Green Revolution. *The Journal of Peasant Studies,* 46(3), 653–673.

Edelman, M. (1980). Agricultural modernization in smallholding areas of Mexico: A case study in the Sierra Norte de Puebla. *Latin American Perspectives,* 7(4), 29–49.

Edelman, M. (2014). Food sovereignty: Forgotten genealogies and future regulatory challenges. *Journal of Peasant Studies,* 41(6), 959–978.

Edwards, P.N. (2010). *A vast machine: Computer models, climate data, and the politics of global warming.* MIT Press.

Escobar, A. (1995). *Encountering development: The making and unmaking of the third world.* Princeton University Press.

Escobar, A. (1998). Whose knowledge, whose nature? Biodiversity, conservation, and the political ecology of social movements. *Journal of Political Ecology,* 5(1), 53–82.

Escobar, A. (2010). Latin America at a crossroads: Alternative modernizations, post-liberalism, or post-development? *Cultural Studies,* 24(1), 1–65.

Escobar, A. (2018). *Designs for the pluriverse.* Duke University Press.

Esteva, G. and Prakash, M.S. (1992). Grassroots resistance to sustainable development: Lessons from the banks of the Narmada. *Ecologist,* 22(2), 45–51.

EZLN (1994). *!Zapatistas! Documents of the New Mexican Revolution.* Autonomedia. Available at: http://lanic.utexas.edu/project/Zapatistas/chapter02.html (last accessed 24 January 2022).

Fairhead, J., Leach, M. and Scoones, I. (2012). Green grabbing: A new appropriation of nature? *Journal of Peasant Studies,* 39(2), 237–261.

Fallon, H., Tollerud, D., Breslow, N., Berlin, J., Bolla, K., Colditz, G., et al (1994). The US military and the herbicide program in Vietnam. In Institute of Medicine, *Veterans and agent orange: Health effects of herbicides used in Vietnam.* National Academy Press, pp 74–110.

Fanon, F. (2001 [1963]). *The wretched of the earth.* Penguin Classics.

Federici, S. (2004). *Caliban and the witch.* Autonomedia.

Federici, S. (2018). *Re-enchanting the world: Feminism and the politics of the commons.* PM Press.

Federici, S. (2019). Social reproduction theory: History, issues and present challenges. *Radical Philosophy,* Spring. Available at: https://www.radicalphilosophy.com/article/social-reproduction-theory-2 (last accessed 16 November 2021).

Fehskens, E.M. (2019). Between the plot and plantation: Parahuman ecologies in Fred D'Aguiar's Children of Paradise. *The Journal of Commonwealth Literature*, 54(2), 124–142.

Feldblum, M. (1998). Reconfiguring citizenship in western Europe. In Joppke, C. (ed) *Challenge to the nation-state: Immigration in western Europe and the US*. Oxford University Press, pp 231–270.

Feltrin, L. and Sacchetto, D. (2021). The work-technology nexus and working-class environmentalism: Workerism versus capitalist noxiousness in Italy's Long 1968. *Theory and Society*, 50, 1–21.

Ferdinand, M. (2021) *Decolonial ecology: Thinking from the Caribbean world*. John Wiley & Sons.

Ferguson, E.B. (1970). The black version of earth day. *The Baltimore Sun*, 14 April, p A10.

Ferretti, F. (2021). Geopolitics of decolonisation: The subaltern diplomacies of Lusophone Africa (1961–1974). *Political Geography*, 85, 102326.

FOEI (2019). A journey through the oil spills of Ogoniland. Friends of the Earth International, 17 May. Available at: https://www.foei.org/a-journey-through-the-oil-spills-of-ogoniland/ (last accessed 22 June 2022).

Forensic Architecture (2021). Environmental racism in death alley, Louisiana. Available at: https://forensic-architecture.org/investigation/environmental-racism-in-death-alley-louisiana (last accessed 22 June 2022).

Forensic Architecture, in partnership with RISE St James (2021). If toxic air is monument to slavery how do we take it down? Available at: https://www.liftfestival.com/digitalevent/if-toxic-air-is-a-monument-to-slavery/ (last accessed 19 December 2022).

Foucault, M. (2007). *Security, territory, population: Lectures at the Collège de France, 1977–78*. Springer.

Frickel, S., Gibbon, S., Howard, J., Kempner, J., Ottinger, G. and Hess, D.J. (2010). Undone science: Charting social movement and civil society challenges to research agenda setting. *Science, Technology, & Human Values*, 35(4), 444–473.

Fukuyama, F. (2006 [1992]). *The end of history and the last man*. Simon & Schuster.

Gabrys, J. (2016). *Program earth*. University of Minnesota Press.

Germain, F. (2008). For the nation and for work: Black activism in Paris of the 1960s. In *Migration and Activism in Europe since 1945*. Palgrave Macmillan, pp 15–32.

Germond, C.S. (2015). An emerging anti-reform green front? Farm interest groups fighting the 'Agriculture 1980' project, 1968–72. *European Review of History: Revue européenne d'histoire*, 22(3), 433–450.

Gibson, N. (2013). Some reflections on Amilcar Cabral's legacy. In Manji, F. and Fletcher, B. (eds) *Claim no easy victories: The legacy of Amilcar Cabral*. Council for the Development of Social Science Research in Africa (CODESRIA), pp 32–40.

Gidwani, V. and Reddy, R.N. (2011). The afterlives of 'waste': Notes from India for a minor history of capitalist surplus. *Antipode*, 43(5), 1625–1658.

Gilio-Whitaker, D. (2019). *As long as grass grows: The Indigenous fight for environmental justice, from colonization to Standing Rock*. Beacon Press.

Gilmartin, M. (2009). Border thinking: Rossport, Shell and the political geographies of a gas pipeline. *Political Geography*, 28(5), 274–282.

Gioielli, R. (2019). Black survival: Mainstream environmentalism's missed opportunities. Enviro-History.com, 22 April. Available at: http://Enviro-History.com/black-survival.html (last accessed 2 June 2022).

Gliessman, S. (2013). Agroecology: Growing the roots of resistance. *Agroecology and Sustainable Food Systems*, 37(1), 19–31.

Glissant, É. (1989). *Caribbean discourse: Selected essays*. Translaed and edited by J. Michael Dash. University Press of Virginia.

Globaïa (2012) *Welcome to the Anthropocene* [film]. Available at: https://globaia.org/anthropocene (last accessed 9 February 2022).

Globaïa (2022) *Vision and mission*. Available at: https://globaia.org/mission (last accessed 9 February 2022).

Gorz, A. (1980 [1975]). *Ecology as politics*. Black Rose Books Ltd.

Gottlieb, R. (2005). *Forcing the spring: The transformation of the American environmental movement*. Island Press.

Gray, R. and Sheikh, S. (2018). The wretched Earth: Botanical conflicts and artistic interventions: Introduction. *Third Text*, 32(2–3), 163–175.

Gudynas, E. (2011). Buen vivir: Today's tomorrow. *Development*, 54(4), 441–447.

Gudynas, E. (2013). Development alternatives in Bolivia: The impulse, the resistance, and the restoration. *NACLA Report on the Americas*, 46(1), 22–26.

Guha, R. (1989). Radical American environmentalism and wilderness preservation: A third world critique. *Environmental Ethics*, 11(1), 71–83.

Guha, R. and Martinez-Alier, J. (2013). *Varieties of environmentalism: Essays north and south*. Routledge.

Gupta, A. (1998). *Postcolonial developments*. Duke University Press.

Hamblin, J.D. (2013). *Arming mother nature: The birth of catastrophic environmentalism*. Oxford University Press.

Haraway, D. (1991). A cyborg manifesto: Science, technology, and socialist feminism in the late twentieth century. In *Simians, cyborgs and women: The reinvention of nature*. Routledge, pp 149–181.

Haraway, D. ([1991] 2020). Situated knowledges: The science question in feminism and the privilege of partial perspective. In *Feminist theory reader*. Routledge, pp 303–310.

Harding, S. (1992). After the neutrality ideal: Science, politics, and 'strong objectivity'. *Social Research*, 59(3), 567–587.

Hare, N. (1970). Black ecology. *The Black Scholar*, 1(6), 2–8.

Harley, J.B., Woodward, D., Edney, M.H., Pedley, M.S. and Monmonier, M.S. (eds) (1987). *The history of cartography*, vol 1. University of Chicago Press.

Harwood, J. (2009). Peasant friendly plant breeding and the early years of the green revolution in Mexico. *Agricultural History*, 83(3), 384–410.

Head, L. (2000). *Second nature: The history and implications of Australia as Aboriginal landscape*. Syracuse University Press.

Held, D. (2005). At the global crossroads: The end of the Washington Consensus and the rise of global social democracy? *Globalizations*, 2(1), 95–113.

Hewitt, D.A. (1976). *Modernizing Mexican agriculture: Socioeconomic implications of technological change 1940–1970*. UN Research Institute for Social Development.

Heynen, N., McCarthy, J., Prudham, S. and Robbins, P. (eds) (2007). *Neoliberal environments: False promises and unnatural consequences*. Routledge.

Higa, T. and Parr, J.F. (1994) *Beneficial and effective microorganisms for a sustainable agriculture and environment*, vol 1. International Nature Farming Research Center.

Holloway, J. (2005). *Change the world without taking power*, expanded edn. Pluto.

Holt-Giménez, E. (2006). *Campesino a campesino: Voices from Latin America's farmer to farmer movement for sustainable agriculture*. Food First Books.

Holt-Giménez, E. and Altieri, M.A. (2013). Agroecology, food sovereignty, and the new green revolution. *Agroecology and Sustainable Food Systems*, 37(1), 90–102.

Humphreys, M. (1996). Kicking a dying dog: DDT and the demise of malaria in the American South, 1942–1950. *Isis*, 87(1), 1–17.

IIED (2014 [1972]). *Survival of spaceship earth*. Available at: https://www.youtube.com/watch?v=KKfBc0Atk8Q (last accessed 12 January 2022).

Indymedia (2005). Irish African unity for justice in Nigeria & Rossport! Available at: http://www.indymedia.ie/article/72983?comment_order=asc&userlanguage=ga&save_prefs=true (last accessed 24 January 2022).

Jackson, M. (ed) (2017). *Coloniality, ontology, and the question of the posthuman*. Routledge.

Jasanoff, S. (2004). Heaven and earth: The politics of environmental images. *Earthly Politics: Local and Global in Environmental Governance*, 31, 41–44.

Jazeel, T. (2005). 'Nature', nationhood and the poetics of meaning in Ruhuna (Yala) National Park, Sri Lanka. *Cultural Geographies*, 12(2), 199–227.

Jessop, B. (2002). Liberalism, neoliberalism, and urban governance: A state-theoretical perspective. *Antipode*, 34(3), 452–472.

Jewett, C. (2017). Recorded talk at 'Thinkery on anti-water privatisation and the commons', University College Cork, 17 June.

Jewett, C. and Garavan, M. (2019). Water is life: An indigenous perspective from a Standing Rock water protector. *Community Development Journal*, 54(1), 42–58.

Jonsson, F.A., Brewer, J., Fromer, N. and Trentmann, F. (2019). *Scarcity in the modern world: History, politics, society and sustainability, 1800–2075*. Bloomsbury Academic.

Karuka, M. (2017). Black and Native visions of self-determination. *Critical Ethnic Studies*, 3(2), 77–98.

Keene, A. and Hitch, G. (2019). *Drone warriors: The art of surveillance and resistance at Standing Rock*. Available at: https://edgeeffects.net/drone-warriors-standing-rock/ (last accessed 9 February 2022).

King, T.L. (2016). The labor of (re) reading plantation landscapes fungible(ly). *Antipode*, 48(4), 1022–1039.

Klein, N. (2001). The unknown icon. *The Guardian*, 3 March. Available at: https://www.theguardian.com/books/2001/mar/03/politics (last accessed 8 July 2021).

Klein, N. (2016). The lesson from Standing Rock: Organizing and resistance can win. *The Nation*, 4 December. Available at: https://www.thenation.com/article/archive/the-lesson-from-standing-rock-organizing-and-resistance-can-win/ (last accessed 24 June 2022).

Kloppenburg, J.R. (2005). *First the seed: The political economy of plant biotechnology*. University of Wisconsin Press.

kollektiv orangotango+ (2018). *This is not an atlas: A global collection of counter-cartographies*. Verlag.

Lakhani, N. (2021). Cop26 legitimacy questioned as groups excluded from crucial talks. *The Guardian*, 8 November. Available at: https://www.theguardian.com/environment/2021/nov/08/cop26-legitimacy-questioned-as-groups-excluded-from-crucial-talks (last accessed 23 November 2022).

Lazure-Beecher, J. (2018). *Nature's empire: Postcolonialism, environmentalism, and Parti pris, 1963–1970*. Doctoral dissertation, Queen's University, Canada.

Lear, L.J. (1998). Rachel Carson's Silent Spring. *Environmental History Review*, 17(2), 23–48.

Leonardi, E. (2012). *Biopolitics of climate change: Carbon commodities, environmental profanations, and the lost innocence of use-value*. Unpublished PhD thesis. Available at: https://ir.lib.uwo.ca/cgi/viewcontent.cgi?article=2239&context=etd (last accessed 18 May 2022).

Leonardi, E. and Benegiamo, M. (2021). André Gorz's labour-based political ecology and its legacy for the twenty-first century. In Räthzel, N., Stevis, D. and Uzzell, D. (eds) *The Palgrave handbook of environmental labour studies*. Palgrave Macmillan, pp 721–741.

Li, T.M. (2002). Engaging simplifications: Community-based resource management, market processes and state agendas in upland Southeast Asia. *World Development*, 30(2), 265–283.

Liboiron, M. (2013). Modern waste as strategy. *Lo Squaderno: Explorations in Space and Society*, 29, 9–12.

Liboiron, M. (2021). *Pollution is colonialism*. Duke University Press.

Liboiron, M. and Lepawsky, J. (2022). *Discard studies: Wasting, systems, and power*. MIT Press.

Lindisfarne, N. (2010). Cochabamba and climate anthropology. *Anthropology Today*, 26(4), 1–3.

Linebaugh, P. (2019). *Red round globe hot burning: A tale at the crossroads of commons and culture, of love and terror, of race and class and of Kate and Ned Despard*. University of California Press.

Lloyd, D. (2003). Rethinking national Marxism: James Connolly and 'Celtic communism'. *Interventions*, 5(3), 345–370.

Loftus, A.J. (2017). Production of nature. In Richardson, D. (ed) *The international encyclopedia of geography: People, the earth, environment, and technology*. John Wiley & Sons, pp 1–6.

Lopez-Carmen, V.A., Erickson, T.B., Escobar, Z., Jensen, A., Cronin, A.E., Nolen, L.T. et al (2022). *US and United Nations pesticide policies: Environmental violence against the Yaqui indigenous nation*. The Lancet Regional Health-Americas.

Lucas Plan (1979). *Lucas Aerospace Combine Shop Stewards Committee Corporate Plan*. Available at: https://www.dropbox.com/s/o2sqxvhams2ywup/Lucas-Plan-53pp-alternative%20corporate%20plan.pdf?dl=0 (last accessed 26 May 2022).

Mahler, A. (2018). *From the tricontinental to the global south: Race, radicalism, and transnational solidarity*. Duke University Press.

Makki, F. (2014). Development by dispossession: Terra nullius and the social-ecology of new enclosures in Ethiopia. *Rural Sociology*, 79(1), 79–103.

Manji, F. and Fletcher, B. (eds) (2013). *Claim no easy victories: The legacy of Amilcar Cabral*. Council for the Development of Social Science Research in Africa.

Manna, J. (2018). *Wild relatives* [film]. 64min, HD video.

Mansfield, B. (2004). Neoliberalism in the oceans: 'Rationalization' property rights, and the commons question. *Geoforum*, 35(3), 313–326.

Mansholt, S. and Walker-Leigh, V. (1974) Interview with Sicco Mansholt. *Ecologist*, 4(7), 259–260.

Marcos, S. (1992). *The southeast in two winds: A storm and a prophecy*. Available at: https://schoolsforchiapas.org/library/southeast-winds-storm-prophecy/ (last accessed 9 July 2021).

Marcos, S. (2011). *Our word is our weapon: Selected writings*. Seven Stories Press.

Marglin, S.A. (1996). Farmers, seedsmen, and scientists: Systems of agriculture and systems of knowledge. In Appfel-Marglin, F. and Marglin, S.A. (eds) *Decolonizing knowledge: From development to dialogue*. Clarendon, pp 185–248.

Martinez-Alier, J. (2003). *The environmentalism of the poor: A study of ecological conflicts and valuation*. Edward Elgar.

Martinez-Alier, J. (2014). 'Growth below zero': In memory of Sicco Mansholt. *EJOLT*. Available at: http://www.ejolt.org/2014/03/growth-below-zero-in-memory-of-sicco-mansholt/ (last accessed 12 January 2022).

Martínez-Torres, M.E. and Rosset, P.M. (2010). La Vía Campesina: The birth and evolution of a transnational social movement. *The Journal of Peasant Studies*, 37(1), 149–175.

Masco, J. (2010). Bad weather: On planetary crisis. *Social Studies of Science*, 40(1), 7–40.

Maximillian, J., Brusseau, M.L., Glenn, E.P. and Matthias, A.D. (2019). Pollution and environmental perturbations in the global system. In Brusseau, M., Pepper, I. and Gerba, C. (eds) *Environmental and pollution science*. Academic Press, pp 457–476.

McAfee, K. (1999). Selling nature to save it? Biodiversity and green developmentalism. *Environment and Planning D: Society and Space*, 17(2), 133–154.

McCarron, M. (2011). Audio Archive. Ken Saro Wiwa Collection, Maynooth University. Available at: https://nuim.libguides.com/ken-saro-wiwa-collection/audio (last accessed 24 January 2022).

McCarthy, J. (2005). Devolution in the woods: Community forestry as hybrid neoliberalism. *Environment and Planning A*, 37(6), 995–1014.

McCord, P.A. (2008). Divergences on the left: The environmentalisms of Rachel Carson and Murray Bookchin. *Left History: An Interdisciplinary Journal of Historical Inquiry and Debate*, 13(1), doi: https://doi.org/10.25071/1913-9632.24606

McCulloch, J. (2019). *In the twilight of revolution: The political theory of Amilcar Cabral*. Routledge.

McGurty, E.M. (1997). From NIMBY to civil rights: The origins of the environmental justice movement. *Environmental History*, 2(3), 301–323.

McKittrick, K. (ed) (2015). *Sylvia Wynter: On being human as praxis*. Duke University Press.

McKittrick, K. (2013). Plantation futures. *Small Axe: A Caribbean Journal of Criticism*, 17(3[42]), 1–15.

McMichael, P. (2014). Historicizing food sovereignty. *Journal of Peasant Studies*, 41(6), 933–957.

McNally, D. (2006). *Another world is possible: Globalization and anti-capitalism*. Arbeiter Ring Publishing.

McVeigh, R. and Rolston, B. (2021). *Ireland, colonialism and the unfinished revolution*. Beyond the Pale Publications.

Meadows, D.H., Meadows, D.L., Randers, J. and Behrens III, W.W. (1972). *The limits to growth: A report for the Club of Rome's project on the predicament of mankind*. Universe Books.

Menchú, R. (1984). *I, Rigoberta Menchú: An Indian woman in Guatemala*. Edited by E. Burgos-Debray and translated by A. Wright. Verso.

Merchant, C. (1981). *The death of nature: Women, ecology, and scientific revolution*. Harper & Row.

Micarelli, G. and Verran, H. (2018). Two analyses of Marisol de la Cadena's Earth beings: Ecologies of practice across Andean worlds. *Tapuya: Latin American Science, Technology and Society*, 1, 123–30.

Midnight Notes (1990). *New Enclosures*, Midnight Notes, 10. Autonomedia. Available at: http://www.midnightnotes.org/newenclos.html (last accessed 26 March 2023).

Millner, N. (2016). Food sovereignty, permaculture and the postcolonial politics of knowledge in El Salvador. In Wilson, M. (ed) *Postcolonialism, indigeneity and struggles for food sovereignty*. Routledge, pp 99–123.

Millner, N. (2017). 'The right to food is nature too': Food justice and everyday environmental expertise in the Salvadoran permaculture movement. *Local Environment*, 22(6), 764–783.

Millner, N. (2020). As the drone flies: Configuring a vertical politics of contestation within forest conservation. *Political Geography*, 80, 102163.

Millner, N. (2021). More-than-human witnessing? The politics and aesthetics of Madre Tierra (Mother Earth) in transnational agrarian movements. *GeoHumanities*, 7(2), 391–414.

Millner, N., Peñagaricano, I., Fernandez, M. and Snook, L.K. (2020). The politics of participation: Negotiating relationships through community forestry in the Maya Biosphere Reserve, Guatemala. *World Development*, 127, 104743.

Mitchell, D. (1996). *The lie of the land: Migrant workers and the California landscape*. University of Minnesota Press.

Monet, J. (2016). Standing Rock joins the world's indigenous fighting for land and life. *Yes Magazine*, 23 September. Available at: https://www.yesmagazine.org/democracy/2016/09/30/standing-rock-joins-the-worlds-indigenous-fighting-for-land-and-life (last accessed 24 June 2022).

Moore, J.W. (ed) (2016). *Anthropocene or capitalocene? Nature, history, and the crisis of capitalism*. PM Press.

Morain, D. (1988). Police batons blamed as UFW official is badly hurt during Bush S.F. protest. *LA Times*, 26 September. Available at: https://www.latimes.com/archives/la-xpm-1988-09-16-mn-2389-story.html (last accessed 25 July 2022).

Morgan, J. and Morgan, N. (2009). *Dr. Seuss and Mr. Geisel: A biography*. Da Capo Press.

Mukherjee, P.U. (2010) *Postcolonial environments: Nature, culture, and the contemporary Indian novel in English*. Palgrave Macmillan.

Murphy, M. (2006). *Sick building syndrome and the problem of uncertainty*. Duke University Press.

Murphy, M. (2008). Chemical regimes of living. *Environmental History*, 13(4), 695–703.

Nash, L. (2004). The fruits of ill-health: Pesticides and workers' bodies in post-World War II California. *Osiris*, 19, 203–219.

Nash, L. (2007). *Inescapable ecologies*. University of California Press.

Nelson, S.H. (2015). Beyond the limits to growth: Ecology and the neoliberal counterrevolution. *Antipode*, 47(2), 461–480.

Niblett, M. (2012). World-economy, world-ecology, world literature. *Green Letters*, 16(1), 15–30.

Nixon, R. (1996). Pipe dreams: Ken Saro-Wiwa, environmental justice, and micro-minority rights. *Black Renaissance*, 1(1), 39.

Nixon, R. (2011). *Slow violence and the environmentalism of the poor*. Harvard University Press.

Nkrumah, K. (1966). *Neo-colonialism: The last stage of imperialism*. International Publishers.

Nunes, R. (2021). *Neither vertical nor horizontal: A theory of political organization*. Verso Books.

Oasa, E.K. and Jennings, B.H. (1982). Science and authority in international agricultural research. *Bulletin of Concerned Asian Scholars*, 14(4), 30–44.

O'Mahony, D. (nd). Airtime. Available at: https://deirdre-omahony.ie/portfolio/airtime/ (last accessed 19 July 2022).

Oreskes, N. (2004). Science and public policy: What's proof got to do with it?. *Environmental Science & Policy*, 7(5), 369–383.

Owens, I.D. (2017). Toward a 'truly indigenous theatre': Sylvia Wynter adapts Federico García Lorca. *Cambridge Journal of Postcolonial Literary Inquiry*, 4(1), 49–67.

Packard, V. (1960). *The waste makers*. David McKay Company.

Panagia, D. (2014). 'Partage du sensible': The distribution of the sensible. In *Jacques Rancière*. Routledge, pp 107–115.

Papadopoulos, D. (2018). *Experimental practice*. Duke University Press.

Papadopoulos, D., Puig de la Bellacasa, M. and Tacchetti, M. (2022a). *Ecological reparation: Repair, remediation and resurgence in social and environmental conflict*. Bristol University Press.

Papadopoulos, D., de la Bellacasa, M.P. and Myers, N. (eds) (2022b). *Reactivating elements: Chemistry, ecology, practice*. Duke University Press.

Paravisini-Gebert, L. (2005). He of the trees. In DeLoughrey, E.M., Gosson, R.K. and Handley, G.B. (eds) *Caribbean literature and the environment: Between nature and culture*. University of Virginia Press, pp 182–198.

Patel, R. (2013). The long green revolution. *The Journal of Peasant Studies*, 40(1), 1–63.

Peet, R. and Watts, M. (2004). *Liberation ecologies: Environment, development and social movements*. Routledge.

Pellow, D.N. (2007). *Resisting global toxics: Transnational movements for environmental justice*. MIT Press.

Perkins, J.H. (2012). *Insects, experts, and the insecticide crisis: The quest for new pest management strategies*. Springer Science & Business Media.

Pimbert, M. (2006). *Transforming knowledge and ways of knowing for food sovereignty*. LIED.

Plumwood, V. (1986). Ecofeminism: An overview and discussion of positions and arguments. *Australasian Journal of Philosophy*, 64(sup1), 120–138.

Plumwood, V. (2002). *Feminism and the mastery of nature*. Routledge.

Prashad, V. (2007). *The darker nations: A biography of the short-lived third world*. Leftword Books.

Puig de la Bellacasa, M. (2010). Ethical doings in naturecultures. *Ethics, Place and Environment*, 13(2), 151–169.

Puig de la Bellacasa, M. (2015). Making time for soil: Technoscientific futurity and the pace of care. *Social Studies of Science*, 45(5), 691–716.

Puig de la Bellacasa, M. (2017). *Matters of care: Speculative ethics in more than human worlds*. University of Minnesota Press.

Pulido, L. (1996). *Environmentalism and economic justice: Two Chicano struggles in the southwest*. University of Arizona Press.

Pulido, L. and Peña, D. (1998). Environmentalism and positionality: The early pesticide campaign of the United Farm Workers' organizing committee, 1965–71. *Race, Gender & Class*, 6(1), 33–50.

Quiroz, M. and Céspedes, C. (2019). Bokashi as an amendment and source of nitrogen in sustainable agricultural systems: A review. *Journal of Soil Science and Plant Nutrition*, 19(1), 237–248.

Radcliffe, S.A. (2012). Development for a postneoliberal era? Sumak kawsay, living well and the limits to decolonisation in Ecuador. *Geoforum*, 43(2), 240–249.

Rancière, J. (1999). *Disagreement: Politics and philosophy*. University of Minnesota Press.

Rancière, J. (2013). *The politics of aesthetics*. Bloomsbury.

Rancière, J. (2015). *Dissensus: On politics and aesthetics*. Bloomsbury.

Rector, J. (2014). Environmental justice at work: The UAW, the war on cancer, and the right to equal protection from toxic hazards in postwar America. *The Journal of American History*, 101(2), 480–502.

Red Nation (2021). *The red deal: Indigenous action to save our earth*. Common Notions.

Reid, J. (2013). Interrogating the neoliberal biopolitics of the sustainable development-resilience nexus. *International Political Sociology*, 7(4), 353–367.

Riofrancos, T. (2019). What green costs. *Logic Magazine*, 9. Available at: https://logicmag.io/nature/what-green-costs/ (last accessed 23 November 2022).

Robertson, M.M. (2006). The nature that capital can see: Science, state, and market in the commodification of ecosystem services. *Environment and Planning D: Society and Space*, 24(3), 367–387.

Rodney, W. (2018 [1972]). *How Europe underdeveloped Africa*. Verso Books.

Rodriguez, A.B. (2015). *Heretical scripts: Sylvia Wynter & the decolonial Atlantic*. Doctoral dissertation, University of Southern California.

Roelvink, G. (2010). Collective action and the politics of affect. *Emotion, Space and Society*, 3(2), 111–118.

Rome, A. (2003). 'Give earth a chance': The environmental movement and the sixties. *The Journal of American History*, 90(2), 525–554.

Rose, D.B. (2004). *Reports from a wild country: Ethics for decolonisation*. UNSW Press.

Rose, D.B. (1996). *Nourishing terrains: Australian Aboriginal views of landscape and wilderness*. Australian Heritage Commission.

Ross, K. (2008). *May '68 and its afterlives*. University of Chicago Press.

Rosset, P.M. and Martínez-Torres, M.E. (2012). Rural social movements and agroecology: Context, theory, and process. *Ecology and Society*, 17(3), 17–29.

Russell, E. (2001). *War and nature: Fighting humans and insects with chemicals from World War I to Silent Spring*. Cambridge University Press.

Sachs, W. (ed) (1997). *The development dictionary: A guide to knowledge as power*. Orient Blackswan.

Said, E. (1978). *Orientalism: Western concepts of the Orient*. Pantheon.

Salleh, A. (2010). From metabolic rift to 'metabolic value': Reflections on environmental sociology and the alternative globalization movement. *Organization & Environment*, 23(2), 205–219.

Sandilands, C. (1997). Mother Earth, the cyborg, and the queer: Ecofeminism and (more) questions of identity. *NWSA Journal*, 9(3), 18–40.

Saro-Wiwa, K. (2012 [1995]). *A month and a day & letters*. Ayebia Clarke Publishing Ltd.

Schiebinger, L. (2004). Feminist history of colonial science. *Hypatia*, 19(1), 233–254.

Schmelzer, M. (2017). 'Born in the corridors of the OECD': The forgotten origins of the Club of Rome, transnational networks, and the 1970s in global history. *Journal of Global History*, 12(1), 26–48.

Schmelzer, M., Vetter, A. and Vansintjan, A. (2022). *The future is degrowth: A guide to a world beyond capitalism*. Verso Books.

Schwarz, C. (2013). Amilcar Cabral: An agronomist before his time. In Manji, F. and Fletcher, B. (eds) *Claim no easy victories: The legacy of Amilcar Cabral*. Council for the Development of Social Science Research in Africa, pp 75–84.

Sheikh, S. (2018). 'Planting seeds/the fires of war': The geopolitics of seed saving in Jumana Manna's wild relatives. *Third Text*, 32(2–3), 200–229.

Simpson, L.B. (2017). *As we have always done: Indigenous freedom through radical resistance*. University of Minnesota Press.

Slobodian, Q. (2008). Dissident guests: Afro-Asian students and transnational activism in the West German protest movement. In *Migration and activism in Europe since 1945*. Palgrave Macmillan, pp 33–55.

Sonnenfeld, D.A. (1992). Mexico's 'green revolution', 1940–1980: Towards an environmental history. *Environmental History Review*, 16(4), 29–52.

Sorensen, B. (2012). Rio+20 demanding accountability. *Cultural Survival Quarterly Magazine*, June. Available at: https://www.culturalsurvival.org/publications/cultural-survival-quarterly/rio20-demanding-accountability (last accessed 22 June 2022).

Spence, M.D. (1999). *Dispossessing the wilderness: Indian removal and the making of the national parks*. Oxford University Press.

Spiro, J. (2009). *Defending the master race: Conservation, eugenics, and the legacy of Madison Grant*. UPNE.

Spivak, G. (1988). Can the subaltern speak? In Nelson, C. and Grossberg, L. (eds) *Marxism and the interpretation of culture*. Macmillan, pp 66–110.

Stoll, M. (2012). *Rachel Carson's Silent Spring: A book that changed the world*. Available at: https://doi.org/10.5282/RCC/3517 (last accessed 25 May 2022).

Sullivan, S. (2013). Banking nature? The spectacular financialisation of environmental conservation. *Antipode*, 45(1), 198–217.

Sullivan, S. (2018a). Making nature investable: From legibility to leverageability in fabricating 'nature' as 'natural capital'. *Science and Technology Studies*, 31(3), 47–76.

Sullivan, S. (2018b). 'Nature 3.0 – will blockchain technology and cryptocurrencies save the planet?' *The Natural Capital Myth and Other Stories*. Available at: https://the-natural-capital-myth.net/2018/02/01/nature-3-0-will-blockchain-technology-and-cryptocurrencies-save-the-planet/ (last accessed 3 February 2022).

Sundberg, J. (1998). NGO landscapes in the Maya biosphere reserve, Guatemala. *Geographical Review*, 88(3), 388–412.

Táíwò, O.O. (2022). *Reconsidering reparations*. Oxford University Press.

Temper, L., Demaria, F., Scheidel, A., Del Bene, D. and Martinez-Alier, J. (2018). The Global Environmental Justice Atlas (EJAtlas): Ecological distribution conflicts as forces for sustainability. *Sustainability Science*, 13(3), 573–584.

Ticktin, M. (2011). The gendered human of humanitarianism: Medicalising and politicising sexual violence. *Gender & History*, 23(2), 250–265.

Todd, Z. (2016). An indigenous feminist's take on the ontological turn: 'Ontology' is just another word for colonialism. *Journal of Historical Sociology*, 29(1), 4–22.

Tronti, M. (1966). 14. Lotta contro il lavoro. In *Operai e capitale*. Giulio Einaudi editore, pp 60–85. [Translated as Tronti, M. [1972] Struggle against labor. *Radical America*, 6(1), 22–25.]

Tsing, A.L. (2011). *Friction: An ethnography of global connection*. Princeton University Press.

Tuck, E. and Yang, K.W. (2012). Decolonization is not a metaphor. *Decolonization: Indigeneity, Education & Society*, 1(1), 1–40.

UNPO (2017). Ogoni. Movement for the survival of the Ogoni people. Member's Profile. Available at: https://unpo.org/downloads/2339.pdf (last accessed 1 February 2022).

Van der Ploeg, J.D. (2012). *The new peasantries: Struggles for autonomy and sustainability in an era of empire and globalization*. Routledge.

Virga, V. (2008). *Cartographia: Mapping civilizations*. Little Brown & Company.

Voyles, T.B. (2015). *Wastelanding: Legacies of uranium mining in Navajo country*. University of Minnesota Press.

Wainwright, J. and Lund, J. (2016). Race, space, and the problem of Guatemala in Miguel Ángel Asturias's early work. *GeoHumanities*, 2(1), 102–118.

Walker, J. and Cooper, M. (2011). Genealogies of resilience: From systems ecology to the political economy of crisis adaptation. *Security Dialogue*, 42(2), 143–160.

Ward, B. and Dubos, R. (1972). *Only one earth: Care and maintenance of a small plant*. W.W. Norton & Company.

Warde, P., Robin, L. and Sörlin, S. (2018). *The environment: A history of the idea*. Johns Hopkins University Press.

Wark, M. (2015). *Molecular red: Theory for the Anthropocene*. Verso Books.

Warren, K. (2000). *Ecofeminist philosophy: A western perspective on what it is and why it matters*. Rowman & Littlefield.

Watts, M. (2001). 1968 and all that… *Progress in Human Geography*, 25(2), 157–188.

Watts, M. (2009). Oil, development, and the politics of the bottom billion. *Macalester International*, 24(1), 79–30.

Watts, M.J. (2013 [1983]). *Silent violence: Food, famine, and peasantry in northern Nigeria*, vol 15. University of Georgia Press.

Wenzel, J. (2017). Turning over a new leaf: Fanonian humanism and environmental justice. In Heise, U.K., Christensen, J. and Niemann, M. (eds) *The Routledge companion to the environmental humanities*. Routledge, pp 181–189.

Wenzel, J. (2019). *The disposition of nature*. Fordham University Press.

Westing, A.H. (1985). The threat of biological warfare. *BioScience*, 35(10), 627–633.

Whatmore, S. (2006). Materialist returns: Practising cultural geography in and for a more-than-human world. *Cultural Geographies*, 13(4), 600–609.

Whatmore, S.J. and Landström, C. (2011). Flood apprentices: An exercise in making things public. *Economy and Society*, 40(4), 582–610.

White, R. (1996). Are you an environmentalist or do you work for a living?: Work and nature. In Cronon, W. (ed) *Uncommon ground: Rethinking the human place in nature*. WW Norton & Company, pp 171–185.

Whyte, K.P. (2016). Why the Native American pipeline resistance in North Dakota is about climate justice. *The Conversation*, 16 September. Available at: https://theconversation.com/why-the-native-american-pipeline-resistance-in-north-dakota-is-about-climate-justice-64714 (last accessed 21 February 2022).

Whyte, K.P. (2017). Our ancestors' dystopia now: Indigenous conservation and the Anthropocene. In Heise, U.K., Christensen, J. and Niemann, M. (eds) *The Routledge companion to the environmental humanities*. Taylor & Francis, pp 208–215.

Wittman, H. (2011). Food sovereignty: A new rights framework for food and nature? *Environment and Society*, 2(1), 87–105.

Woelfle-Erskine, C. (2019). Beavers as commoners? Invitations to river restoration work in a beavery mode. *Community Development Journal*, 54(1), 100–118.

Wood, D. (2016). Imbrications of coloniality. An introduction to Cabralist critical theory in relation to contemporary struggles. In Rabaka, R. and Cabral, A. (ed) *Resistance and decolonization*. Rowman & Littlefield, pp 43–72.

Wynter, S. (1971). Novel and history, plot and plantation. *Savacou*, 5, 95–102.

Wynter, S. (1994 [1992]). No humans involved: An open letter to my colleagues. *Forum NHI: Knowledge for the 21st Century*, 1(1), 42–73.

Wynter, S. and McKittrick, K. (2015). Unparalleled catastrophe for our species? Or, to give humanness a different future: Conversations. In *Sylvia Wynter: On being human as praxis*. Duke University Press, pp 9–89.

Ybarra, M. (2018). *Green wars: Conservation and decolonization in the Maya forest*. University of California Press.

Yusoff, K. (2017). Epochal aesthetics: Affectual infrastructures of the anthropocene. *E-flux*, 29 March. Available at: https://www.e-flux.com/architecture/accumulation/121847/epochal-aesthetics-affectual-infrastructures-of-the-anthropocene/ (last accessed 24 June 2022).

Yusoff, K. (2018). *A billion black Anthropocenes or none*. University of Minnesota Press.

Ziegler, J., Golay, C., Mahon, C. and Way, S. (2011). *The fight for the right to food: Lessons learned*. Springer.

Index

Page numbers in *italic* type refer to figures. References to endnotes show both the page number and the note number (145n2).

A

aesthetics 8, 154n6
   aesthetic regimes 5, 10, 131, 145n2
Africa
   afro-futurism 109
   anti-colonial movements 56
   decolonisation 11–12, 66–69
'Africa Europe Faith and Justice Network' 93
African National Congress 152n21
African Wildlife Foundation 18
'afterlives' of social movements 143
Agarwal, A. 11
Agent Orange 31, 45
Agent White 31
agroecology 4, 13, 26, 38, 39, 41–42, 67, 113, 133, 140, 151n10
agronomy, and traditional agriculture 40–41
ALBA (Bolivian Alliance for the Peoples of our Americas) 108
Algerian War of Independence (1954–1962) 57
Allen, Will 49
alter-globalisation 94, 95, 98, 99, 108, 143
Alves, Maria Theresa 125
al-Wardi, Umar bin Muzaffar Ibn 77, *77*
Amazon, the 58–59
Amnesty International 91, 92
Anchorage Declaration (Indigenous Peoples' Global Summit on Climate Change, 2009) 155n15
Anders, William 5, 54
Anthropocene 105, 106, 107, 112–113, 117, 121–123
(anti-)colonial geographies 8–12
anti-colonial movements 2, 10, 12, 14, 26, 55, 56–57, 58–59, 61, 66–72, 88–89, 129, 130–131, 137
anti-imperialist movements 13, 26, 59, 133
anti-nuclear movement 62–63
anti-racism movements 2, 69, 129, 130
anti-war movements 55

Anzaldúa, Gloria 152
apartheid, South Africa 152n21
Arawak peoples 70
Astier, M. 40
Asturias, Miguel Ángel 71
Attenborough, David 105
Australia, British colonisation of 10
Ausungate earth being 155n7

B

Bandung Conference, Indonesia, 1955 58
Barca, Stefania 14, 63, 106
Bennholdt-Thomsen, Veronika 14
Beothuk people 125
Bhopal gas tragedy, India 46, *46*, 47
Biafran War 90
BINGOs (big NGOs) 146n15
Biosphere Reserves 146n7
Black ecology 36–37
Blackfeet people, US 9
Blaikie, Piers 13
BLM (Black Lives Matter) movement 128
'Blue Marble' image *78*, 79
bokashi 114–115, 116
Bolivia 109, 154n4
Bolivian Alliance for the Peoples of our Americas (ALBA) 108
Bonneuil, Christophe 8
Bookchin, Murray 29–30
Borlaug, Norman 38, 39
Borneo 51
Boudia, Soraya 8
Brandt Line 151n9
Braun, Bruce 88
Bretton Woods Institutions 83
Bristol, UK
   Colston statue 128
   'Seeds of change' floating ballast seed garden, 2012–2016 (Alves) 125

INDEX

B

Brown, Freddie Mae 36
Brundtland Commission Report, 1987 (UN) 82
'*buen vivir*' (living well) 154n4

C

Cabral, Amílcar 12, 13, 40, 43, 57, 66–69, 70, 86
California
  chemical pesticides and migrant workers 14–15, 27, 30, 32–33, 34–36, 132, 138
  consumer boycott of grapes 36, 132
Campesino-a-Campesino (farmer-to-farmer) movement 114, 151n10
Cape Verde 66, 67, 68
Captain Planet and the Planeteers 81, 82
Carib peoples 70
Caribbean 11–12, 56, 69–72
Carson, Rachel 5, 25–26, 27, 41, 42, 49, 50, 133
  context of 28–31
Casement, Roger 153n23
Central America 85, 87
Césaire, Aimé 43, 59, 68, 69
Césaire, Suzanne 43
Chávez, César 30, 33
chemical fertilizers 38, 116
chemical herbicides, Vietnam War 31, 45
chemical manufacture 62
  St James Parish, Louisiana 128, 129–130, 139–140
chemical pesticides 5, 25–26, 31, 116, 137–138, 147n1
  California 14–15, 27, 30, 32–33, 34–36, 132, 138
  and the Green Revolution 38
  industrial workers and environmental health 32–37
  Mexico 39
  promotion of 49, *49*
  resistance to 29–30
  and warfare 26–27, 30–31, 49, 149n15
  see also Silent spring (Carson)
chemical warfare 31, 45
Chevron 146n15
Chiapas, Mexico 87, 94, 95–99, 139
China, Cultural Revolution (1966–1976) 57
Christian, Dorothy 120
citizen science 5, 35
Civil Rights Movement 55, 65
CLEAR (Civic Laboratory for Environmental Action Research), Newfoundland 125
Cleaver, Harry Jr. 65, 153n28
Climate Camp, Kerry, Ireland 142
climate change 11, 72, 107, 122
CLIMAVORE 126–127
Club of Rome 57, 148n3, 150n1
CO_2 (carbon) emissions 17–18

Cochamba, Bolivia, WPCCC (World People's Conference on Climate Change), 2010 107–111
CODECA (Comité de Desarrollo Campesino /Campesino Committee of Development) 103
Cold War 12, 28, 31, 55, 80, 82
colonialism 10, 16, 121–122, 140, 142
  anti-colonial movements 2, 10, 12, 14, 26, 55, 56–57, 58–59, 61, 66–72, 88–89, 129, 130–131, 137
  and chemical pesticides 26, 27
  neocolonialism 55, 58–59
Colston, Edward 128
Comité de Desarrollo Campesino / Campesino Committee of Development (CODECA) 103
'Common Ecologies' 156n1
commoning 19, 86
commons 84
composting practices 114–115
Congo 153n23
Connolly, James 19–20
Connolly, William 139
Conservation International 18
conservationism 9–10
consumer boycott of Californian grapes 36, 132
COP-15 Summit 107
COP-26 (United Nations Climate Change Conference, Conference of the Parties), Glasgow 2021 1, 18, 146n16
'Copenhagen Accord,' 2009 107
Cosgrove, Denis 7, 75, 78–79
Costa Rica 114, 117
critical race theory 146n9
Cronon, William 7, 9
Cruz Jimenez, Antonio 153n2
Cultural Revolution, China (1966–1976) 57

D

Dalla Costa, Mariarosa 63–64, 65
DAPL (Dakota Access Pipeline) 117–119
Davis, H. 106
DDT (dichlorodiphenyltrichoroethan) 17–18, 27, 29, 30–31, 33, 35, 42, 45, 50, 149n15
  commercial promotion of *48*, 48–49
  environmental legislation, US 25–26, 30, 39, 132
De Castro, José 58–59, 67
De la Cadena, Marisol 109–110
decolonisation 4, 11–12, 13, 56, 59, 72, 73, 142
  Africa 11–12, 66–69
  Caribbean 69–72
  decolonial environmental politics 109, 133
  of science 121
  US 37
degrowth 150n20
Deloria, Vine Jr. 37

177

disagreement 23, 41, 107, 117, 122–123, 132, 134–135, 137, 140, 143
Don't Look Up 1, 2
Dorsey, M.K. 146n15
Dow Chemical 31, 46
Downstream (Christian and Wong) 120
drones 120
DuPont 45, 46, 49, 146n15

E

earth beings 12, 100, 101, 107, 109–110, 112, 123, 135
Earth Day, 1970 36
earth politics 4, 19, 105–107, 121–123, 132, 134, 140
  ICARDA (International Centre for Agricultural Research in Dry Areas) seed conservation 111–112
  permaculture practices in El Salvador 112–117
  Standing Rock water protectors 107, 117–121, 122, 132, 142, 156n19
  WPCCC (World People's Conference on Climate Change), Bolivia, 2010 107–111
Earth Summit (UN Conference on Environment and Development), Rio de Janeiro, 1992 82, 84, 151n8
Earth systems science 106, 154n2
'Earthrise' image 5, 54, 55, 78, 78–79, 106
echo chamber 138–139
ecofeminism 13, 14, 64, 88, 134, 140, 151n13
Ecologist, The, 'Whose common planet?' issue 85
Ecuador 109, 154n4
Edwards, Paul 8
Ehrlich, Paul 60
El Inca Garcilaso de la Vega 79–80
El Salvador, permaculture 112–117
electric power, Guatemala 103
enclosures 83–85, 132
Energuate 103
environment, the, as a concept 6
'environmental colonialism' 11
environmental conservation 17–18
environmental governance 82, 84–85, 132
environmental health 4
  chemical pesticides and industrial workers 32–37, 39
environmental justice 12, 14, 134, 147n17
  and environmental racism 36–37
  St James Parish, Louisiana 128, 129–130, 139–140
environmental management 131–132
environmental movements 16
  post-1968 period 25, 58–59, 131–132, 137
Environmental Protection Agency, US 30, 36
environmental racism 36–37

environmentalism
  1968–1973 period 56
  aesthetics of 134
  denaturalising of 7–8
  early environmentalism, geographies of 25–28, 42–43
    context of 28–31
    industrial workers, pesticides and environmental health 32–37
    Mexico, and the Green Revolution 37–42
  meaning of 2
  provincialising of 15–16, 134
erasure practices, European colonialism 10, 11, 12
Escobar, A. 110
eugenics 9, 145n5
Europe
  agricultural reform and the 'Mansholt Plan' 60
  colonialism and erasure practices 10
Experimental Farm of Pessubé 67
expert knowledge 18
  see also science
ExxonMobil 146n15
EZLN (Zapatista National Liberation Army) 95, 96, 97, 141

F

FA (Forensic Architecture) 128, 129–130, 139–140
Fanon, Franz 12, 36–37, 43, 59, 64, 66, 68, 71
farmer-to-farmer (Campesino-a-Campesino) model 114, 151n10
Federici, Silvia 63, 65, 146n114
Feltrin, L. 62
feminism 146n9
  ecofeminism 13, 14, 64, 88, 134, 140, 151n13
  post-1968 movements 56, 59, 60, 63–65, 73, 137
  second wave 28
'fence-line' communities 14
  St James Parish, Louisiana 128, 129–130, 139–140
'field,' as a geographical space for agricultural technology experimentation 26
First World War 29, 30
Flit 49, 49
flooding, climate-induced 17
Floyd, George 128, 129
food security 85, 86
food sovereignty 4, 26, 67–68, 86, 87, 97, 134, 140
Forensic Architecture (FA) 128, 129–130, 139–140
'forever chemicals' 46
Foucault, M. 151n7

INDEX

Fressoz, Jean-Baptiste 8
Friends of the Earth UK 92
'fruit culture,' California 32, 36
Fukuyama, Francis 150n2

G

Gabrys, Jennifer 8
Geisel, Theodore (Dr Seuss) 49
gender, and labour 14
  see also social reproduction
General Motors Corporation 149n10
Geneva Protocol (1925) 31
gentrification 99
Gilio-Whitaker, Dina 155n14
Glacier National Park, US 9
Glissant, Édouard 43, 70
Globaïa 105–106, 108–109, 110, 121
Global South 11, 15
globalisation 135
  alter-globalisation 94, 95, 98, 99, 108, 143
'god-trick' 17
gold mining, the Sperrins, Ireland 142
Goldman Prize 94
Gorz, André 59, 60–61, 73–74
governance, environmental 82, 84–85, 132
governmentality, green 151n7
Grant, Madison 9
green governmentality 151n7
green investment 84
'green' products 61
Green Revolution 27, 42, 43, 85, 111, 113, 116, 140
  Japan 155n10
  Mexico 26, 27, 37–42
green technologies 11, 84
greenhouse gas emissions, scientific accounting of 11
Greenpeace 52, 91, 92
Guamán Poma de Ayala, Felipe 80
Guatemala 152n15
  Civil War 89
  'witnessing' 102–104
Guha, Rabindranath 16, 133
Guinea-Bissau 66, 67

H

Haraway, Donna 7, 17
Hardin, Garret 60
Harding, Sandra 7
Hare, Nathan 36–37
Hargreaves, Malcolm 29
Hawk's Nest Tunnel disaster, West Virginia 47
Health and Safety Clause 35–36
Hecht, Gabrielle 8
herbicides *see* chemical herbicides, Vietnam War
Higa, Professor 155n10
Hitch, G. 120

Holling, C.S. 145n3
Huerta, Dolores 30, 33
humanitarianism 88

I

ICARDA (International Centre for Agricultural Research in Dry Areas) seed conservation 111–112
iconography 99–100
  planetary icons 75, 76, 77, 77–80, 78, 79, 80
  see also Madre Tierra (Mother Earth)
Illich, Ivan 60
imperialism
  anti-imperialist movements 13, 26, 59, 133
Inca cosmology 79, 79–80
incommensurability 109, 138
India, Green Revolution 26, 39, 42, 43
Indigenous Americans 9
  educational work 155n14
  resistance to settler-colonialism 37, 118, 133
  Standing Rock water protectors 107, 117–121, 122, 132, 142, 156n19
Indigenous futurism 109
Indigenous knowledge/science 15, 120–121
Indigenous land 4, 11, 16, 37, 122, 140
  post-1968 movements 55, 66–72
  water rights 1–7, 117–121, 122, 132, 142
Indigenous ontologies 19, 109, 110, 118
Indigenous people 13
  Chiapas, Mexico 39, 87, 94, 95–99, 139
  COP-26 1
  land dispossession 9, 11, 12
  resistance to settler-colonialism in the US 37, 118, 133
Indigenous Peoples' Global Summit on Climate Change, 2009 155n15
Indigenous sovereignty 4, 117, 118, 120, 122, 134, 140
Indonesia 152n18
industrial technology, critiques of 62–63
industrial workers, chemical pesticides and environmental health 32–37
information and communication technologies 6, 138–139
INGOs (international non-governmental organisations) 17, 18
Institute of Agricultural Research (Instituto de Investigaciones Agricolas), Mexico 41
Intergovernmental Panel on Climate Change (IPCC) 6, 11, 81, 82
International Centre for Agricultural Research in Dry Areas (ICARDA) seed conservation 111–112
International Monetary Fund 83, 153n29
international non-governmental organisations (INGOs) 17, 18

International Union for the Conservation of Nature
  World Conservation Congress 146n16
International Wages for Housework Campaign 63–64, 65
internet 6
  see also information and communication technologies
IPCC (Intergovernmental Panel on Climate Change) 6, 11, 81, 82
Ireland 21, 146n12
  anti-gold-mining campaign, the Sperrins 142
  Republican independence struggle 19–20
  resistance to Shell and support for Ogoni people 53, *93*, 93–94
  Standing Rock water protectors 119, 122
  'Making Relatives' events, 2022 142
  T.U.R.F. (Transitional Understandings of Rural Futures), 2013 (O'Mahony) 125
  Zapatista 'Journey for Life' visit, 2021 141–142
Istanbul, Turkey, CLIMAVORE 126

J

James, Selma 63–64
Japan 114, 117, 155n10
Jas, Nathalie 8
Jewett, Chas 118–120, 122, 156n19

K

Keene, A. 120
Kerry, Ireland 142
'Kharidat al-'Aja'ib wa Faridat al-Ghara'ib' Persian cosmology map 77, *77*
Klein, Naomi 98
Knobloch, Frieda 7
knowledge
  alternative infrastructures of 18–19
  Indigenous knowledge/science 15, 120–121
  politics of 109
  see also expert knowledge; science

L

La Via Campesina (LVC) 83, 85–89, 98, 99, 100, 120, 132, 135
'laboratory,' as a geographical space for agricultural technology experimentation 26
labour
  labour/social reproduction blindspot 8, 12–16
  post-1968 movements 55, 56, 57, 59, 60, 61–63, 64–65, 73
land *see* Indigenous land
'Land in Our Names' 156n1
Latin America 85, 87, 108, 118, 120
  Green Revolution 37, 39

Lazure-Beecher, J. 59
Leonardi, E. 63
'liberation ecology' 13
liberation theology 95–96, 114
Liboiron, Max 8, 121
Limits to growth report (OECD), 1972 6, 60, 148n3, 150n1
Limón, Edmundo 40
Linares, Enrique 153n2
Linebaugh, Peter 20, 65
lived ecologies and everyday experiments blindspot 8, 17–19
living well ('*buen vivir*') 154n4
Los Angeles, US, CLIMAVORE 126
Lucas Aerospace Corporation UK 62–63
LVC (La Via Campesina) 83, 85–89, 98, 99, 100, 120, 132, 135

M

Maastricht Treaty 153n28
Maathai, Wangari 153n29
madre (mother) 116
Madre Tierra (Mother Earth) 75, 83, 87–88, 89, 98, 99, 100–101, 102, 103, 108, 113, 114, 115, 116–117, 123, 132, 134, 135
magical realism 71
mainstreaming 99, 132
making things resonate *see* resonance
Malaya 31
Manna, Jumanna 111–112
Mansholt, Sicco 59–60, 61
Mao Zedong 57
MAP (Mexican Agricultural Program) 37–42
maps 75, 76
  'Kharidat al-'Aja'ib wa Faridat al-Ghara'ib' Persian cosmology map 77, *77*
Marcos, S. 95, 99
Marcuse, Herbert 60
Martinez-Alier, Juan 16, 66, 133
mass consumption 61
Mayan culture 87, 116, 151n11
McAfee, K. 18
McCarron, Majella 53, 92–94, *93*
McKittrick, K. 71
Médecins sans Frontières 88
Memmi, Albert 59
Menchú, Rigoberta 89
Metropolitan Black Survival Committee 36
Mexican Agricultural Program (MAP) 37–42
Mexico
  Constitutional amendment allowing the privatisation of communal land 153n25
  Green Revolution 26, 27, 37–42, 43
  oil industry 51
  Zapatista movement 16, 53, 87, 89, 91, 94, 95–99, 100, 120, 132, 135, 139
    'Journey for Life' visit to Ireland 141–142
Midnight Notes Collective 83, 84, 85
Mies, Maria 14

migrant workers
  chemical pesticides 26
  California 14–15, 27, 30, 32–33, 34–36, 132, 138
Mi'kmaq people 125
Mitchell, Don 32–33
Mitman, Gregg 8
#Mniwiconi 155n13
modern environmentalism 1–4
  (anti-)colonial geographies blindspot 8–12
  history of 4–8
  labour/social reproduction blindspot 8, 12–16
  lived ecologies and everyday experiments blindspot 8, 17–19
Monsanto 116, 146n15
Morales, Evo 107
Morales, Jimmy 103
MOSOP (Movement for the Survival of the Ogoni people) 53, 89, 91–94, *93*, 97, 98, 99, 100, 120, 132, 135
Mother Earth *(Madre Tierra)* 75, 83, 87–88, 89, 98, 99, 100–101, 102, 103, 108, 113, 114, 115, 116–117, 123, 132, 134, 135
Mukherjee, Pablo 66
Murphy, Michelle 8
Murphy, Robert Cushman 29
music, and resonance 136

N

NAFTA (North American Free Trade Agreement) 95, 139
NAM (Non-Aligned Movement) 58, 80, *80*, 151n9
Narain, S. 11
Nash, Linda 8, 33, 34, 35
National Farm Workers Association 33
National Parks 9
nature
  queering of 88
Nature Conservancy 18
nature, Romantic perspectives and wilderness imaginaries 7, 9, 10, 133
Nelson, Sara 145n3
neocolonialism 55, 58–59
neoliberalism 18, 83, 96, 153n28
net-zero targets 1
Newfoundland, Canada, CLEAR (Civic Laboratory for Environmental Action Research) 125
Niblett, M. 71
Niger Delta, oil extraction 16, 51–53, *52*, *53*, 75, 87, 89–94, *93*, *94*, 97, 132, 135
Nigeria 51–53, 152n16
  see also Niger Delta, oil extraction
Nixon, Rob 17–18, 27, 92, 152n17
no-growth economy 60
nuclear energy/weapons
  anti-nuclear movement 62–63

nuclear testing 5, 43
nuclear war 5
Nunes, Rodrigo 156n4

O

Office of Special Studies (OSS) 41
Ogoni people 16, *52*, 52–53, *53*, 75, 87, 89–94, *93*, 97, 132, 135
Ojibwe Ceremonial Water Teachings 119
O'Mahony, Deirdre 126
'one-worldism' 54–57, 72, 87, 107, 134
'ontological turn' 109
organophosphates 33, 34, 35–36
OSS (Office of Special Studies) 41
overfishing 12
overpopulation 13

P

Pachamama 108, 113, 123
Pan-Indianism 37
Panzieri, Raniero 65
Papadopoulos, Dimitris 18–19, 121, 156n17
Paris, France, civil unrest, 1968 56, 57, 58, 135
participatory approaches 151n8
  environmental governance 84–85
Partido Africano da Independencia da Guine e Cabo Verde 67
Partis Pris 59
PCBs (polychlorinated biphenyls) 36
permaculture 132, 134
  El Salvador 112–117
Persian cosmology 77, *77*
pesticides *see* chemical pesticides
PFAs (perflouroalkyl/polyflouroalkyl) 46
Pharand-Deschênes, Félix 105
Pinchot, Gifford 145n5
Planetary Boundaries project, Stockholm Resilience Centre 105
planetary icons 75, *76*, 77, 77–80, *78*, *79*, *80*
plantation, and plot 69–72
plastics 44, 45
  PVC 62
  St James Parish, Louisiana 128, 129–130, 139–140
plot, and plantation 69–72
pluriverse 109–110, 123, 130, 132, 134, 135, 137
political ontology 101, 110, 118
Porto Maghera, Italy 62, 63
postcolonialism 40, 43
Prashad, Vijay 56–57
preservationism 9–10
pristine nature 7, 9
'Psalter Map, The' *76*
Puig de la Bellacasa, Maria 113–114, 147n18, 156n16
PVC plastics, manufacture of 62

Q

Q'eqchi' Maya people 87, 103
Québec, Canada 59
Quechua cosmology 108
queering of nature 88

R

racism
   anti-racism movements 2, 69, 129, 130
   and chemical pesticides 26, 27, 30, 34
   environmental 36–37
   and labour 14
   racial spatial segregation 36–37
   and *terra nullius* principle 10
   toxic waste dumps 147n17
Rancière, Jacques 110, 134, 139, 145n2, 154n6
Reagan, Ronald 153n28
Red Power 37
reparation 56, 71, 72, 130, 134, 142, 149n14
resilience, as a concept 145n3
resonance 104, 124–127, 129, 130–131, 135–139, 140, 143
Rights for Nature 118, 120
RISE St James 128, 129–130, 139–140
Rockefeller Foundation 38, 40
Rodney, Walter 148n6
Rodriguez, A.B. 69
Romanticism, and nature 7, 9, 10
Roosevelt, Theodore 145n5–6
Rose, Deborah Bird 14
Ross, Kristin 57, 143
Rossport, Ireland, Shell to Sea campaign 53, *53*, 94
Royal Dutch/Shell *51*, 51–53, *52*, *53*, 89–94, *93*, 99
Ruiz Garcia, Samuel 95–96
Runakuna people 155n7
Russell, E. 29

S

Sacchetto, D. 62
sacrifice zones 71
Salinas de Gortari, Carlos 139, 153n25
Salvadorean Civil War (1979–1992) 113
Santa Cruz Pachacuti, Juan de *79*, 79–80
Saro-Wiwa, Ken 16, 53, 75, 89, 90–93, *93*, 94, 96, 99, 100, 132, 135
Sartre, Jean-Paul 60
Sauer, Carl 40
scenes 8
science 17
   decolonisation of 121
   and Indigenous knowledge 15, 120–121
   'white science' 39
   see also expert knowledge
Seattle, US, street protests, 1999 99, 143
Second World War 29, 30, 49, 145n1

seeds
   genetically modified 38, 39, 116 (*see also* Green Revolution)
   ICARDA (International Centre for Agricultural Research in Dry Areas) seed conservation 111–112
   'Seeds of change' floating ballast seed garden, 2012–2016 (Alves) 125
   seed-saving practices 40, 111, 112
Senghor, Léopold Sédar 59
Shell Oil Company *see* Royal Dutch/Shell
Shell to Sea campaign, Ireland 53, *53*, 94
'sick-building' syndrome 8
Silent spring (Carson) 5, 25–26, 27, 41, 42, 49, 50, 133
   context of 28–31
Sioux Tribe, Standing Rock water protectors 107, 117–121, 122, 132, 142, 156n19
Skye, Scotland, *CLIMAVORE* 126–127
Slater, Candace 7
slavery 64, 69, 70, 89, 129
   Bristol 128
   historic public memorials 128, 139
'slow violence' 14, 27, 92
social movements 3–4
   'afterlives' of 143
   anti-colonial movements 2, 10, 12, 14, 26, 55, 56–57, 58–59, 61, 66–72, 88–89, 129, 130–131, 137
   anti-imperialist movements 13, 26, 59, 133
   anti-nuclear movement 62–63
   anti-racism movements 2, 69, 129, 130
   anti-war movements 55
   environmental movements 16
   post-1968 period 25, 58–59, 131–132, 137
social reproduction 14, 28
   post-1968 movements 56, 59, 60, 63–65, 73
social theory, provincialising of 15–16
soil 12, 67, 68, 113
   and permaculture practices 113–117
South Africa, boycott of goods 152n21
South-East Asia 11–12
space travel 5, 55
   see also 'Blue Marble' image; 'Earthrise' image
'spaceship earth' 54, 55
Sperrins, the, Ireland 142
Spivak, Gayatri 15
St James Parish, Louisiana 128, 129–130, 139–140
'stakeholding,' and environmental governance 84–85
Standard Oil Company 49
Standing Rock water protectors 107, 117–121, 122, 132, 156n19
   'Making Relatives' events, Ireland, 2022 142

INDEX

Starr, Kevin 32
Steinbeck, John 147n4
'stewardship' 84
Stockholm Resilience Centre 105
Structural Adjustment Programmes 83
student unrest, 1968 56, 57, 58, 135
Subaltern Studies Group 15
'suburb,' as a geographical space
  for agricultural technology
  experimentation 26
sugar plantations 69–70
Sullivan, Sian 18, 154n3
Sumak Kawsay ('*buen vivir*'/living well) 154n4
Summers, Laurence 146n10
Survival International 92
sustainable development 12, 81–82, 84, 132, 134, 151n8
Svalbard Global Seed vault 111

T

Taboada, Edmundo 40, 41
Táíwò, Olúfémi O. 149n14
Teflon 46
terra nullius ('land belonging to no one') 10–11, 17, 133
testimonio (witnessing) 88–89
Thatcherisation 153n28
Third World 12, 56–57, 135
  emergence of as a concept 57–58
'Third-Worldism' 56, 57–58, 66, 71–74
Thomas, Wilbur 36
Todd, Z. 106
toxic waste dumps 11, 14, 36, 133, 140, 147n17
Transitional Understandings of Rural Futures (T.U.R.F.), 2013 (O'Mahony) 125
Tronti, Mario 61–62, 64
Tsing, Anna 100, 146n13, 152n18
Tuck, E. 122
T.U.R.F. (Transitional Understandings of Rural Futures), 2013 (O'Mahony) 125

U

UFW (United Farm Workers) 33, 34–35, 36
UN (United Nations)
  Agenda 21 82
  Brundtland Commission Report, 1987 82
  Climate Change Conference, Conference of the Parties (COP-26), Glasgow 2021 1, 18, 146n16
  Conference on the Human Environment, Stockholm, 1972 54, 82
  COP summits 11
  Earth Summit (UN Conference on Environment and Development), Rio de Janeiro, 1992 82, 84, 151n8
  Framework Convention on Climate Change 107
  Rio+20 Earth Summit, 2012 105, 121
United Nations Educational, Scientific and Cultural Organisation
  Biosphere Reserves 146n7
United Nations Environmental Program 54
  Conference, 1972 55
UNPO (UN Working Group on Indigenous Populations and Unrepresented Nations and People's Organisation) 153n22
World Food Summit, 1996 86
Uncommon ground 7
Union Carbide 46–47, 47
United Automobile Workers of America 149n10
United Farm Workers (UFW) 33, 34–35, 36
Universal Declaration of the Rights of Mother Earth, The 154n4
US Agency for International Development 38
US (United States)
  as birthplace of modern environmentalism 7–8
  Cold War geopolitics 28
  decolonisation 37
  environmental justice 37
  St James Parish, Louisiana 128, 129–130, 139–140
  environmental legislation
  banning of DDT 25–26, 30, 39, 132
  toxic waste dumps 147n17
  see also California; Indigenous Americans

V

vampire metaphor 96
Vietnam War (1954–1975) 57
  herbicide use 31, 45

W

Wallace, Henry 37, 39
Ward, Barbara 5, 55
warfare
  and chemical pesticides 26–27, 30–31, 49, 149n15
  chemical warfare 31, 45
Warren County, North Carolina 36
Washington Consensus 84, 108
waste dumps, toxic 11, 14, 36, 133, 140, 147n17
water, and the commons
  Standing Rock water protectors 107, 117–121, 122, 132, 142, 156n19
Watts, Michael 13, 57, 58, 89
Welcome to the Anthropocene film (Globaïa) 105–106, 108–109, 110
Wenzel, Jennifer 66, 137
Wheatland Riot, 1913 32
White, Richard 7, 13
'white science' 39
Wild Relatives (Manna) 111–112

183

wilderness imaginaries 7, 9, 10, 133
wildfires, North America 128
Wildlife Conservation Society 18
Wilson, Waziyatawin Angela 118, 119
wind 155n11
'witnessing' 17, 88–89, 100, 102–104, 120, 139
Wiwa, Ownes 153n24
women *see* feminism; social reproduction
Wong, Rita 120
work *see* labour
worker-scientists 35
working class 14, 64, 65
'working-class ecology' 63
World Bank 38, 83, 153n29
World People's Conference on Climate Change (WPCCC), Cochamba, Bolivia, 2010 107–111
'world problematique' 57–59, 65

World Social Forum 108
World Wide Fund for Nature 18
Worster, Donald 7
WPCCC (World People's Conference on Climate Change), Cochamba, Bolivia, 2010 107–111
Wynter, Sylvia 12, 57, 68, 69–70, 71–72

Y

Yang, K.W. 122
Yusoff, K. 146n9

Z

Zapata, Emiliano 95
Zapatista movement, Mexico 16, 53, 87, 89, 91, 94, 95–99, 100, 120, 132, 135, 139
   'Journey for Life' visit to Ireland 141–142
Zerowork collective 64–66, 150n3